HEREDITY AND DEVELOPMENT

Heredity and

Development

JOHN A. MOORE

PROFESSOR OF ZOOLOGY, BARNARD COLLEGE
AND COLUMBIA UNIVERSITY

NEW YORK OXFORD UNIVERSITY PRESS 1963

1 3 5 6

Preface

AN INDIVIDUAL IS A PRODUCT of his heredity and his development. His heredity is the substance he receives from his parents—his biological inheritance. The essence of this substance is the set of instructions that it contains. An ovum and a sperm, the hereditary substance of man, unite to form a fertilized ovum, the zygote. The zygote contains all the instructions required to produce another human being.

Shortly after its formation, the zygote undergoes a series of changes that leads, if its luck holds, to an adult individual. These changes are its development. Development is species-specific, that is, the sequence of changes from zygote to adult is controlled by the nature of the hereditary instructions. For one zygote the instructions may be "make a man," for another "make a mouse." If all goes well, one will in due time be a man, the other a mouse. Development, then, may be thought of as a carrying out of the hereditary instructions contained in the zygote.

The problems of heredity and development have been central to biology since this field began, in the mid-nineteenth century, to become a rigorous science. The problems of heredity and development will always be central to biology. This is inevitable since they are closely associated with the unique feature of life itself—the ability to replicate.

This book is concerned with heredity and development. Its basic plan is to show how ideas in these two fields were first formulated and then studied. The intellectual history of the two has been quite different. Genetics, the science of heredity, effectively began in 1900. By 1930 the general laws of inheritance seemed to be well established, in the sense that the rules for the transmission of genes from parent to offspring were understood. Furthermore the rules, with only slight modifications, applied to all animals and plants that geneticists studied.

Classical genetics was essentially complete by 1930. It was in a state similar to that of physics in 1899, when A. A. Michelson said,

The more important fundamental laws and facts of physical science have all been discovered, and these are now so firmly established that the possibility of their ever being supplanted in consequence of new discoveries is exceedingly remote. . . . "Our future discoveries must be looked for in the sixth place of decimals."

For both sciences the period of intellectual calm was brief. Physics was soon revolutionized by studies of the nucleus of physical matter; genetics was revolutionized by further studies of the nucleus of living matter. In the decade beginning with 1944, geneticists discovered the chemical substances of which genes are composed. With this as a basis they have gone on to show how the chemical structure of the gene is responsible for its specific activity.

It is hard to exaggerate the importance of these discoveries. So far as biology is concerned, they are matched in importance only by the theory of evolution. For science in general they represent one of the great intellectual achievements of the twentieth century.

Embryology, the science of development, has not had such a progressive history. It had a vigorous beginning and the main events and problems were soon demarcated. But explanation has come hard and the fundamental causes of events in development are still poorly understood. It was inevitable that this be so. There could be no satisfactory explanations in embryology until there was a reasonable comprehension of the action of genes. Now that we seem close to such an understanding, this is more likely. Embryology is a science of tomorrow.

The science of biology has few stories more interesting than the history of man's attempts to explain heredity and development.

Chapters 1 through 16 and 19 through 21 are reprinted, almost without change, from the author's *Principles of Zoology* (Oxford University Press, New York, 1957). Chapters 17 and 18 contain new materials on genetics, beginning with transformation in Diplococcus and ending with 'cracking the code' of DNA. In Chapter 22 an attempt is made to synthesize the explanatory hypotheses of genetics and embryology. Betty Moore and F. J. Ryan have kindly checked the new chapters for me. The original line drawings were prepared from my rough sketches by Russell Peterson and Frank Romano.

Contents

THE DEVELOPMENT OF GENETIC CONCEPTS

PROSPECTUS

Some of the most fundamental biological events are taken for granted. We 'naturally expect' the offspring of human beings to be other human beings; of pine trees to be pine trees; of Amoeba to be Amoeba. But a scientist cannot 'naturally expect' anything when there is the possibility of his discovering causal relations. Natural events must have as their basis laws that are either known or knowable. At least, a belief of this sort is the working philosophy of scientists.

Establishing the laws that explain why it is that offspring resemble their parents has been one of the most exciting chapters in modern biology. A bewildering mass of observations has been united in a conceptual scheme that is rational and concise. This is the science of genetics.

In all probability men have speculated about the nature of heredity since the dawn of history. It can be said safely, however, that their speculations led to nothing that could be called science until quite recently.

Our method of approach to the field of genetics will be a historical one. We shall begin with the year 1868, when Charles Darwin published *The Variation of Animals and Plants under Domestication,* and trace the increase in man's understanding of genetic phenomena down to the present day. This method of approach will illustrate how a science develops.

It is possible that this approach will reveal some things about science that you never realized before. We are so accustomed to a succession of triumphs in science that we fail to realize that this is not an inevitable consequence of the application of the 'scientific method.' Perhaps you have some knowledge of the way scientific understanding is thought to come about. The explanation is usually given as follows:

1. A scientist is confronted with a natural phenomenon that he wishes to explain.

2. He invents a hypothesis to explain the unknown phenomenon in terms of known phenomena.
3. Deductions are made from the hypothesis.
4. The deductions are tested. If they are found to be true it is probable that the hypothesis is true. The more deductions that are tested and found to be true the more probable it is that the hypothesis is true.
5. Eventually the scientist convinces himself and his contemporaries that his hypothesis adequately explains the phenomenon.
6. In the course of explaining the first unknown phenomenon other unknown phenomena will be encountered. The same procedure is followed to explain these and so successive phenomena become understandable and organized into the ever-growing body of scientific knowledge.

At least that is the way that science is thought to work! To the non-scientist it may appear that if one has the equipment, time, intelligence, and abides by the rules, scientific progress is inevitable. To those working in science, this view is clearly a distortion. Progress is not always along a straight line. In our survey of genetics, for example, we shall find that many blind alleys have been entered and in addition we shall learn that an 'established fact' of one period may turn out to be an 'erroneous conclusion' in the light of later discoveries. The idea of straight line progress is a distortion due to looking back on the history of science. When we do this we select the experiments, observations, and ideas that subsequent events have shown to be correct and we ignore the inadequate experiments, incorrect observations, and faulty ideas.

We shall attempt, therefore, to describe the progress of genetics as it took place. We shall try to understand some of the problems facing scientists who are working on unknown phenomena. It is most important that we make the attempt; we are living in an age when scientific knowledge is of the utmost concern to all mankind. The proper use of scientific knowledge can result in unparalleled benefits to mankind and a misuse can lead to unimaginable disasters. It is essential that those who will make the decisions, and they will be the non-scientists primarily, have as much knowledge of the nature of scientific methods as possible.

Our purpose, therefore, in studying genetics is twofold. We should gain some appreciation of the data and concepts in the field of genetics and, in addition, an understanding of the manner in which science develops.

1

Darwin's Theory of Pangenesis—1868

BACKGROUND FOR THE THEORY OF PANGENESIS

Charles Darwin's interest in genetics was a consequence of his studies of evolution. It will be necessary, therefore, to give a brief statement of his evolution theory in order to show its relation to genetics.

Darwin imagined that evolution occurred in this manner: Among the individuals of any species there would be many differences. For example, some might be slightly larger than the average, or have longer legs, or have a thicker coat of fur. If any of these variations made their possessors better adapted to survive, those with the better characteristics would have a greater chance of leaving offspring ('survival of the fittest,' as Spencer later described it). With the passage of time the original population would change, its individuals gradually becoming larger, or developing longer legs or a thicker coat of fur, or whatever characteristic was of value for survival. In this way one species could evolve into another or give rise to two or more different species.

We cannot at this time discuss in detail Darwin's theory of evolution. For the present, we should merely note the importance of variations. Evolution cannot occur unless there are differences among the individuals of the same species. If all individuals are identical and remain so generation after generation, obviously there is no evolution. So variation is essential and, furthermore, to be of importance in evolution it must be inherited. A thick coat of fur might be advantageous for a mammal living in the Arctic, but unless this variation is inherited it is unimportant for evolution.

Darwin fully realized that his theory of evolution must be based on a sound understanding of the mechanism of inheritance. In an attempt to provide such a basis, he collected all the data possible on animal and plant breeding, and then developed the first comprehensive theory of heredity, or as we now call it, genetics. This appeared in 1868 as a two-volume work entitled *The Variation of Animals and Plants under Domestication*. In this he assembled many observations on inheritance, largely of domestic plants and animals, and attempted to provide a theory to explain these observations.

Darwin's work in this field was of major interest in the last half of the nineteenth century. He was, of course, the outstanding biologist of his time, so anything he did attracted attention. In addition to this, for many years his theory was the only one available. We shall examine his theory briefly, not only for its historical interest but to see how the problems were stated, what data were available, and finally, to what extent the theory contributed to an understanding of natural events.

Let us constantly keep in mind that our purpose is twofold: first, learning genetics, and second, learning how scientific theories develop and change. For this second purpose, it will be important for us to keep an open mind and, if possible, not to be prejudiced by what we may have read or learned before. It is difficult not to be influenced by what we may know of genetics, but if possible this knowledge should be ignored. If we are discussing the state of genetics in 1868 our approach should be this: Given the data available in 1868, how would we view the problems of inheritance?

First, something of the background for Darwin's work will be given. Knowledge of cell structure, which in later years was to form a foundation for genetic concepts, was in a rudimentary state. It was known that animals and plants were composed of cells, but little was known about the internal structure of cells. The nucleus was thought to be a universal cell constituent, although its role in the life of the cell was unknown. It was generally believed, as we believe today, that cells arise solely from pre-existing cells and not *de novo*. Opinions on heredity were vague and varied. The crossing of varieties in both animals and plants had been practiced for centuries, but no general laws or rules to explain the results had been discovered. In fact, the data were so confusing that some doubted that they could be scientifically explained.

One type of observation that convinced Darwin of the 'force of inheritance' was that 'with man and the domestic animals, certain peculiarities have appeared in an individual, at rare intervals, or only one or twice in the history of the world, but have reappeared in several of the children or grandchildren.' One of the most spectacular instances of this was the porcupine man, whose skin was covered by warty projections (Fig. 1–1). Six of his children and two of his grandchildren showed this same defect. Another instance was found in some domestic

Phw. Trans. Vol. XLIX. TAB. I p. 21.

1–1 The hand of the porcupine man.

pigs, entirely lacking hind legs, whose abnormality was carried through three generations.

To most biologists of the mid-nineteenth century, such instances seemed to be the result of mere chance, or of environmental influence, but to Darwin they were evidence that 'something' was transmitted from parent to offspring. The following quotation illustrates the way he reasoned.

When we reflect that certain extraordinary peculiarities have thus appeared in a single individual out of many millions, all exposed in the same country to the same general conditions of life, and again, that the same extraordinary peculiarity has sometimes appeared in individuals living under widely different conditions of life, we are driven to conclude that such peculiarites are not directly due to the action of the surrounding conditions, but to unknown laws acting on the organisation or constitution of the individual;—that their production stands in hardly closer relation to the conditions than does life itself. If this be so, and the occurrence of the same unusual character in the child and parent cannot be attributed to both having been exposed to the same unusual conditions, then the following problem is worth consideration, as showing that the result cannot be due, as some authors have supposed, to mere coincidence, but must be consequent on the members of the same family inheriting something in common in their constitution. Let it be assumed that, in a large population, a particular affection occurs on an average in one out of a million, so that the *a priori* chance that an individual taken at random will be so affected is only one in a million. Let the population consist of sixty millions, composed, we will assume, of ten million families, each containing six members. On these data, Professor Stokes has calculated for me that the odds will be no less than 8333 millions to 1 that in the ten million families there will not be even a single family in which one parent and two children will be affected by the peculiarity in question. But numerous cases could be given, in which several children have been affected by the same rare peculiarity with one of their parents; and in this case, more especially if the grandchildren be included in the calculation, the odds against mere coincidence become something prodigious, almost beyond enumeration.

Even today, it would be hard to supply better reasons for the belief that 'something' was transmitted from the first porcupine man to his son. Darwin ruled out the possibility of the external environment having any causal relation to the appearance of the defect: If something in the environment was the stimulus, why did just these few persons and no others have the defect? Surely if there was some unusual feature of the environment, such as rare climatic conditions or a peculiar substance in the diet, many individuals might be expected to have a 'porcupine skin.' Neither could it be due to chance. It was inconceivable that a defect, so rare as never to be recorded before, should affect a father, son, and grandson merely by chance.

Darwin concluded that the best explanation was that the son had inherited his father's defect. This, in turn, means that something is transmitted from father to son. If this is the case, it should be possible to obtain information on the laws governing this transmission. If these laws could be formulated, not only would this represent a tremendous advance for genetics, but a firm foundation would be provided for the theory of evolution.

The Data To Be Explained. Darwin's procedure, that is, his 'scientific method,' was as follows: First, he assembled all the information he could find that seemed to have a bearing on heredity. Second, he proposed a theory to account for all of the information he had assembled. The mass of data contained in his two-volume work is considerable, but it can be combined into a small number of categories. These were the types of data that Darwin felt must be explained by any comprehensive theory of inheritance.

1. Transmission of Characters from Parent to Offspring. Darwin summarized a tremendous mass of observations on this topic. Most of the characters known to him were morphological, such as differences in body size, type of feathers, or hair and color patterns. Others were physiological; examples are the inheritance of profuse bleeding in man, and peculiar tics or nervous defects. The inherited characters might be large or small, important or unimportant. He concluded, 'When a new character arises, whatever its nature may be, it generally tends to be inherited, at least in a temporary and sometimes in a most persistent manner.' It is clear that Darwin had no conception of an orderly or predictable transmission of characters from parent to offspring. Inheritance to him was a capricious phenomenon, sometimes temporary and sometimes persistent.

2. Mutilations. Some races of man habitually knock out their teeth, cut off parts of their fingers, or perforate their ears or nostrils, yet their children do not show corresponding defects. There were other cases where mutilations appeared to be inherited and they were given on such good authority that Darwin found it 'difficult not to believe them.' One of these was 'a cow that had lost a horn from an accident with consequent suppuration, produced three calves which were hornless on the same side of the head.' Once again the situation was complex. Mutilations appeared to be inherited in some instances but not in others.

3. Atavism (or Reversion). This is the presence in an individual of some peculiar characteristic not expressed in its immediate parents, but resembling a remote ancestral condition. Children occasionally resemble their grandparents or more remote ancestors more closely than they do their parents. Domestic animals may have peculiar features not characteristic of their breed, but resembling the wild species from which the domestic forms were derived. Black sheep (it was thought that sheep in early times were dark) occasionally arise in carefully bred flocks of white sheep. Instances are reported of reversion during the life of a single individual. Darwin crossed white hens with black cocks.

Some of the chicks were white the first year and the same individuals became black the second.

4. *Sex.* For most characters it appeared that inheritance could be by way of the male or the female, but Darwin knew of a few instances in which the sex of the parent was important in inheritance. Cases are quoted on traits being transmitted from father to son but never to daughters, or from mother to daughter but never to sons. In color blindness, males are much more commonly affected than females, yet the defect can be transmitted through normal females. In fact, it seemed probable to Darwin that fathers can never transmit color blindness to their sons. Daughters of color-blind fathers, on the other hand, though normal themselves, transmit color blindness to their sons. 'Thus, the father, grandson, and great-great-grandson will exhibit a peculiarity—the grandmother, daughters, and great-granddaughter having transmitted it in a latent state.' From observation of this sort Darwin states, 'We thus learn, and the fact is an important one, that transmission and development are distinct powers.'

5. *Inbreeding and Inheritance.* If two organisms are crossed and their offspring bred with each other generation after generation, we speak of this as inbreeding. The data available to Darwin suggested that if one began with two types of organisms and inbred them generation after generation, there would result a relatively homogeneous population in which there is a blending of characteristics. Darwin regarded this as the general rule, but he adds (in small print!) that in other cases 'some characters refuse to blend, and are transmitted in an unmodified state either from both parents or from one. When gray and white mice are paired, the young are not piebald nor of an intermediate tint, but are pure white or the ordinary gray color.'

6. *Selection and Inheritance.* Selection is a breeding method that has been employed since the early days of agriculture. If a farmer is interested in increasing the size of his chickens, he selects the largest individuals and breeds them. From their offspring he selects the largest and breeds from them. With this procedure, it is usually possible to increase the average size of the descendants in a few generations. One of the most puzzling aspects of selection was the fact that frequently it was possible to produce an organism with characteristics not even remotely suggested in the original stock. For example, by continued selection it was possible to produce the most bizarre varieties of pigeons that had characteristics that did not occur in the ancestors. In short, selection could create something new. This will be considered in the following section.

7. Origin of Variability. All domestic and wild species familiar to Darwin were variable. Numerous varieties of roses or pigeons were known, for example. Many varieties bred true, indicating the hereditary nature of the special features. In some cases a variety was known to have originated from a single exceptional individual. In many cases it appeared that the new variety was 'new' in the sense of never having occurred before. The factors concerned with the origin of variability were important and had to be considered in any comprehensive theory of inheritance. The cause of variability was 'an obscure one; but it may be useful to probe our ignorance.' Darwin favored the view that 'variations of all kinds and degrees are directly or indirectly caused by the conditions of life to which each being, and more especially its ancestors, have been exposed.' The great importance of the 'conditions of life' can be brought out by the following quotation: '. . . if it were possible to expose all the individuals of a species during many generations to absolutely uniform conditions of life, there would be no variability,' The actual conditions of life that were thought to cause variability included excess food (probably the most important), climate, hybridization, grafting in plants, and in fact 'a change of almost any kind in the conditions of life.'

8. Regeneration. When the tail or the legs of a salamander are cut off the lost structures are replaced perfectly by regeneration. The regenerated leg or tail is a *salamander* leg or tail. The ability to regenerate lost parts is of widespread occurrence and appears to be similar to events occurring in embryonic development. Darwin felt that both the formation of a structure during the course of normal development and its replacement following injury to the adult were due to the workings of inheritance.

9. Inheritance and Mode of Reproduction. There are two main types of reproduction, sexual and asexual. An animal like Hydra is capable of both. Sexual reproduction consists of the fertilization of an ovum by a spermatozoon. Asexual reproduction in Hydra is by budding. A small protuberance forms on the side of the Hydra. This grows and eventually detaches as a small individual. The Hydra that originates from a fertilized ovum is identical with a Hydra developing from a bud. Inheritance is the same whether by sexual or asexual means, in Darwin's opinion.

10. Delayed-Action Inheritance. Darwin listed several cases, which he believed to be well substantiated, of the male gametes having an effect on the female organs. One of these was published by Lord Morton. An Arabian chestnut mare was crossed to a quagga (a wild African species belonging to the horse genus and closely resembling the zebra). One offspring was obtained and it was intermediate in form and color. The mare was subsequently crossed to a black Arabian horse. One filly and one colt were produced. In coloration and type of mane these two offspring showed a striking resemblance to the quagga. For example, dark bars were present on the hind part of the body and the mane was stiff and erect. Darwin concluded, 'Hence, there can be no doubt that the quagga affected the character of the offspring subsequently begot by the black Arabian horse.' He felt that the quagga sperm had acted directly on the reproductive organs of the female in such a way as to affect the characteristics of future offspring sired by other males.

THE THEORY OF PANGENESIS

These ten categories represent the types of data that Darwin felt must be explained by any comprehensive theory of inheritance. After collecting as much of the pertinent data as he could, he set about to formulate a theory. The result was '. . . the hypothesis of Pangenesis, which implies that the whole organization, in the sense of every separate atom or unit, reproduces itself.' The starting point of this is the postulation of gemmules that determine all characteristics of the organism. The properties that gemmules were assumed to possess were these: Each and every cell of an organism, and even parts of cells, produce gemmules of a specific type corresponding to the cell or part. These are able to circulate throughout the body and they become aggregated in the sex cells. Every sperm and every egg will contain gemmules of all sorts. By the union of egg and sperm they are transmitted to the next generation. During development they unite with partially formed cells or with other gemmules, and in this way produce new cells of the type from which they were formed. In some instances the gemmules could remain dormant for generations. We should think of a liver cell as producing gemmules for every part of that cell, enough kinds to produce the identical cell type in the next generation. All other parts of the body would also be producing their own specific gemmules. These must be present in tremendous numbers, since every sperm and ovum will have some of all types produced in the body.

Today we might wonder about the space problem. If every part of the body produced a specific gemmule, would it not be difficult for all of them to fit into an ovum of microscopic dimensions, or into the even smaller sperm? If gemmules exist, obviously they must be very small. Darwin did not think this difficulty was fatal to his hypothesis. Biologists at that time, and especially those working on disease or with cells, realized that very small things could be extremely important.

The questions of basic importance for his theory concerned the existence of the gemmules and their production by cells. Was there any evidence for their existence? At the time Darwin wrote we must remember that the 'cell theory' was in the process of being accepted. Darwin reasoned this way: If cells can divide and produce other cells, perhaps they can produce other bodies with the assumed characteristics of gemmules by a similar process. In his own words, 'The existence of free gemmules is a gratuitous assumption, yet can hardly be considered as very improbable, seeing that cells have the power of multiplication through the self-division of their contents.' Darwin's whole Theory of Pangenesis depended on the actual existence of gemmules, and it should be realized that he had no direct observational evidence for them. In short, he invented them to account for observed data in the field of heredity.

This is legitimate scientific procedure. Atoms were invented to account for the data of chemistry. The planet which was later named Pluto was invented to account for certain irregularities in the orbits of known planets. But to postulate is not to prove. The facts available to Darwin were not sufficient to decide whether his theory was 'right' or 'wrong.' His main contribution was the collection of a tremendous amount of genetic data, and an attempt to provide a theoretical framework for its interpretation. He was most modest about his efforts: 'I am aware that my view is merely a provisional hypothesis or speculation; but until a better one be advanced, it may be serviceable by bringing together a multitude of facts which are at present left disconnected by any efficient cause. As Whewell, the historian of the inductive sciences, remarks: "Hypotheses may often be of service to science, when they involve a certain portion of incompleteness, and even of error." Under this point of view I venture to advance the hypothesis of Pangenesis, which implies that the whole organization, in the sense of every separate atom or unit, reproduces itself.'

The Theory Explains the Data. Let us now apply the Theory of Pangenesis to the ten categories of data requiring explanation.

1. Transmission of Characters from Parent to Offspring. The appearance of the same characters in parent and offspring was made possible by the production of gemmules by all parts of the parent's body. These entered the ova and sperm and by fertilization were transmitted to the offspring where they caused their specific effects. This was true, as well, for those special characters such as those of the porcupine man.

The skin cells of the porcupine man produced 'porcupine' gemmules. These reached his children by way of the sperm.

2. Mutilations. Mutilations are usually not inherited because the part to be mutilated would have produced gemmules before its removal. These gemmules would enter the gametes and be passed to the next generation. The few cases in which mutilations appeared to be inherited usually involved diseased parts. Darwin explained this as follows: 'In this case it may be conjectured that the gemmules of the lost part were gradually all attracted by the partially diseased surface, and thus perished.'

3. Atavism. Atavism, according to the Theory of Pangenesis, was due to the ancestral gemmules remaining in a dormant condition for many generations and then suddenly developing.

4. Sex. Both sexes transmit inherited characters with equal facility, since both transmit gemmules representing every cell of the body. In the case of color blindness in man, and similar instances of inheritance modified by sex, it was assumed that gemmules were latent in one sex. A color-blind man transmits gemmules of color blindness to his daughter (in whose body they are dormant) and she may in turn transmit them to her sons. In the sons they develop and the sons are color-blind.

5. Inbreeding and Inheritance. The blending in the offspring of characteristics of the parents is due to the mixing of the gemmules of the parental types. Those cases in which the characteristics of one parent predominate merely indicate that the predominating ones 'have some advantage in number, affinity, or vigour over those derived from the other parent.'

6. Selection and Inheritance. It is possible to influence the inherited characteristics of organisms through selection in this manner: The farmer choosing the largest chickens from his flock is choosing the ones that will produce gemmules for large size. If this is repeated every generation, the gemmules for small size will be eliminated and the chickens reach their maximum possible size.

7. Origin of Variability. According to Darwin, new characteristics appear as a result of some environmental influence. The new or changed structure will produce new types of gemmules. These will be transmitted to the next generation, and thus the new character will reappear.

8. Regeneration. Regeneration of lost parts is possible because the gemmules for the lost parts were produced prior to the loss and are present in the rest of the body. If, for example, the leg of a salamander has been removed, the leg gemmules, which are present in the body, can migrate to the cut surface and develop into a new limb, identical to the old one.

9. Inheritance and Mode of Reproduction. Inheritance is the same, whether via sexual or asexual means, since the basis is identical—the transmission of gemmules. In the case of our specific example, Hydra, every cell of the body would produce gemmules. These would move to all parts, including the gametes and the cells that form the buds. Thus, the new individual would receive the same gemmules irrespective of whether they came from a fertilized ovum or from a bud.

10. Delayed-Action Inheritance. In those peculiar cases where the male gametes were thought to have a lasting effect on the reproductive organs of the females (as in Lord Morton's mare) a ready explanation was possible. Some of the gemmules from the male gametes entered the reproductive organs of the female and were included in ova produced long afterwards.

Darwin's Theory of Pangenesis, like all great theories, involved a great simplification in man's view of his universe. By assuming the existence of gemmules with definite properties he was able to 'make sense' out of a previously bewildering mass of data. Inheritance was not a nebulous and capricious force, but a precise and orderly transmission of the physical entities that are the basis of development in succeeding generations.

We should now pause to ask a few questions: Did Darwin's Theory of Pangenesis explain the data of heredity? If you are aware of later developments in this field you will probably answer 'no,' but if you can repress the bias of the knowledge of what was to come, you will probably conclude that the answer is 'yes.' If the answer is 'yes,' does this mean that the theory is correct?

Darwin's approach to the study of inheritance was one of two possible methods of attack. He was concerned nearly entirely with the *results* of inheritance, i.e. the kind of offspring obtained when parents of different types were crossed. From the results he attempted to reconstruct the basis of inheritance. As we have already seen, he concluded that every cell produces gemmules and that these are the basis of inheritance. In the thirty years following the presentation of the Theory of Pangenesis little or no advance, based on breeding experiments, was made in our understanding of the mechanism of inheritance.

The second possible method of investigation involves the study of ova and sperm. These gametes are the sole physical link between the parents and offspring, so presumably they would be responsible for the transmission of any inherited characteristics. A careful study of the gametes might be expected to throw some light on inheritance. The branch of biology that is concerned with the study of cells, including the ova and sperm of course, is *cytology*. It was in the field of cytology that the major advances in understanding inheritance were made during the last half of the nineteenth century. In the next few chapters we shall consider these findings.

SUGGESTED READINGS

Darwin, C. 1868. *The Variation of Animals and Plants under Domestication.* Murray. This work has been reprinted in many editions. Chapter 27 is devoted to the 'Provisional Hypothesis of Pangenesis.'

The Cell and Its Division

The development of cytology, like that of all fields of science, depends partly on tools. The compound microscope, which is the basic tool in cytology, appears to have been invented about 1590 by two Dutch spectacles makers. The birthday of cytology was postponed, however, for three-quarters of a century. It was not until 1667 that Hooke first described 'cells' in a piece of cork (Fig. 2–1). This discovery of cells in cork could have been an important advance in knowledge or an unimportant one. If it had turned out subsequently that cells were found only in cork, Hooke would certainly not be widely remembered for his discovery. But the work of many scientists showed that cells were a general phenomenon and, therefore, of some importance. We might conclude that Hooke did not make a discovery that *was* important but, instead, a discovery that *became* important.

The Cell Theory. Following the realization that some cells are present in some organisms, the next conceptual advance was to establish the fact that all organisms are composed solely of cells or cell products. This took nearly a century and a half. In the early years of the nineteenth century a number of cytologists came more and more to adopt this second concept, which we know as the 'cell theory.'

Three names are generally associated with the final formulation of the cell theory. They are Dutrochet, Schleiden, and Schwann. The last-named published his treatise on the microscopic structure of organisms in 1839, when he was 29 years old. In it he summarized his findings on a variety of tissues. His general conclusion was that the bodies of organisms are composed of cells. Figure 2–2 reproduces some of his drawings.

Obſerv. XVIII. *Of the* Schematiſme *or* Texture *of* Cork, *and of the Cells and Pores of ſome other ſuch frothy Bodies.*

I Took a good clear piece of Cork, and with a Pen-knife ſharpen'd as keen as a Razor, I cut a piece of it off, and thereby left the ſurface of it exceeding ſmooth, then examining it very diligently with a *Microſcope*, me thought I could perceive it to appear a little porous; but I could not ſo plainly diſtinguiſh them, as to be ſure that they were pores, much leſs what Figure they were of: But judging from the lightneſs and yielding quality of the Cork, that certainly the texture could not be ſo curious, but that poſſibly, if I could uſe ſome further diligence, I might find it to be diſcernable with a *Microſcope*, I with the ſame ſharp Pen-knife, cut off from the former ſmooth ſurface an exceeding thin piece of it, and placing it on a black objeſt Plate, becauſe it was it ſelf a white body, and caſting the light on it with a deep *plano-convex Glaſs*, I could exceeding plainly perceive it to be all perforated and porous, much like a Honey-comb, but that the pores of it were not regular; yet it was not unlike a Honey-comb in theſe particulars.

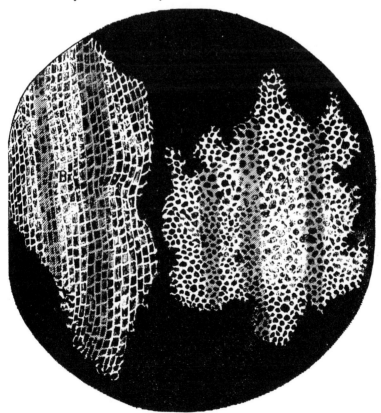

2–1 The drawings and part of the text from Hooke's observations on cork (R. Hooke, *Micrographia*, London, 1667).

17

Onion Cells

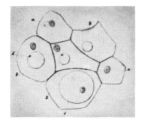

Notochord of Fish

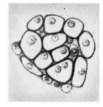

Cartilage of Frog

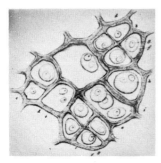

Cartilage of Tadpole

Muscles of Foetal Pig

Areolar Tissue of
Pig Embryo

Capillary in
Tadpole Tail

Ganglion Cells
of Frog

Cells of Pig
Embryo

2-2 Some of Schwann's drawings of cells (Th. Schwann, *Mikroscopische Unter-*
suchungen über die Uebereinstimmung in der Struktur und dem Wachsthum der
Thiere und Pflanzen, Berlin, 1839).

The following quotations from a translation of Schwann's treatise
will reveal some of his views:

Though the variety in the external structure of plants is great, their internal
structure is very simple. This extraordinary range of form is due only to a
variation in the fitting together of elementary structures which, indeed, are
subject to modification but are essentially identical—that is, they are cells.
The entire class of cellular plants is composed solely of cells which can readily
be recognized as such; some of them are composed merely of a series of similar
or even only of a single cell.

Animals being subject to a much greater range of variation in their external
form than is found in plants also show (especially in the higher species) a

much greater range of structure in their different tissues. A muscle differs greatly from a nerve, the latter from a cellular tissue (which shares only its name with the cellular tissue of plants), or elastic tissue, or horny tissue, etc. If, however, we go back to the development of these tissues, then it will appear that all these many forms of tissue are constituted solely of cells that are quite analogous to plant cells . . . . The purpose of the present treatise is to prove the foregoing by observations.

Much of Schwann's success was due to the fact that he adopted a definite criterion for the recognition of cells, namely, the presence or absence of a nucleus. The latter structure was apparently first recognized as an important and characteristic cell structure by Robert Brown in 1831. It had been observed much earlier, however. Schwann took full advantage of this recent (for him) discovery.

The most frequent and important basis for recognizing the existence of a cell is the presence or absence of the nucleus. Its sharp outline and its darker color make it easily recognizable in most cases and its characteristic shape, especially if it contains nucleoli . . . identify the structure as a cell nucleus and make it analogous with the nucleus of the young cells contained in cartilage and plant cells. . . . More than nine-tenths of the structures thought to be cells show such a nucleus and in many of these a distinct cell membrane can be made out and in most it is more or less distinct. Under these circumstances it is perhaps permissible to conclude that in those spheres where no cell membrane be distinguished, but where a nucleus characteristic of its position and form is encountered, that a cell membrane is actually present but invisible.

Although Schwann's work established the cell as a unit of structure, his views on the origin of cells precluded these elementary bodies from having any importance in inheritance, as the following quotation shows.

The general principles in the formation of cells may be given as follows. At first there is a structureless substance which may be either quite liquid or more or less gelatinous. This, depending on its chemical constitution and degree of vitality, has the inherent ability to bring about the formation of cells. It seems that usually the nucleus is formed first and then the cell around it. Cell formation is in the organic world what crystallization represents in the inorganic world. The cell, once formed, grows through its inherent energy, but in doing so it is guided by the organism as a whole in the way that conforms to the general organization. This is the phenomenon basic to all animal and plant growth. It is applicable to cases where the young cells originate in the mother cell, as well as those where they are formed outside of them. In both instances the origin of cells occurs in a liquid or in a structureless substance. We will call this substance, in which cells are formed, a cell germinative substance or Cytoblastema. It can be compared figuratively, but only figuratively, with a solution from which crystals are precipitated.

The Continuity of Cells. Gradually this view was replaced by the realization that cells are formed solely by the division of pre-existing cells. Even before 1839 cell division had been observed, but it was during the following decade that more and more investigators—Remak and Nägeli, for example—came to the conclusion that cells never originate from a structureless 'cytoblastema,' but always by cell division. In 1855 Virchow formulated his well-known statement *omnis cellula e cellula,* which means 'all cells from cells.' The cell then took on a new significance and greater importance. No longer was it a matter of the organism forming cells *de novo,* but instead cells formed the organism. If existing cells have arisen from pre-existing cells, then there must be a continuity of these elementary structures that goes back to the very beginnings of life. The connection between generations was shown to be cells. Schwann was of the belief that the ovum was a cell. Therefore a cell produced in the ovary was the link between parent and offspring. (It had been known since 1824 that spermatozoa, and not the fluid in which they are found, were the important agents in fertilization, but the realization that spermatozoa are cells did not come until 1865.)

Nuclear Division. With the gradual accumulation of knowledge, improvement in microscopes, and development of techniques, cytologists were able to see more and more detail in cells. The nucleus, being the most characteristic structure within the cell, came in for a good deal of attention. Its role in cell division was not understood at first. Some observers held that the nucleus disappears during cell division and each daughter cell produces a new one. Using this interpretation, there is no connection between the nuclei of different cell generations. Others believed that during cell division the nucleus was constricted and pinched in two and then one part went into each daughter cell. The first of these beliefs was held mainly by botanists and the second by zoologists. It goes without saying that the nucleus would be without importance in the transmission of hereditary factors if its existence terminated at each cell division.

It should be kept in mind that when living cells are examined it is frequently difficult to see the nucleus. During cell division this rather indistinct body does disappear, especially if one is looking through a crude microscope. It appears as though a new nucleus forms in each of the daughter cells. The difficulty in making observations, coupled with the fact that methods of fixation and staining were poorly developed, makes it easy to understand why cytologists believed what their eyes told them: that the nucleus disappears during cell division. Nevertheless there were many others who *believed* that the nucleus was not completely dissolved, but that in some manner a portion of the original nucleus gave rise to the daughter nuclei.

In the year 1873 three biologists independently described complex nuclear changes which occurred during cell division and which are now termed *mitosis*. (Simultaneous, though independent, discovery is common in science. We shall have more examples of it.) They were Schneider, Bütschli, and Fol. Schneider's paper appeared first (a 'paper' is an article appearing in a scientific journal). It was not concerned with cell division, but with the morphology of a flatworm called Mesostoma. The bulk of the paper is taken up with details of the structure but, being a careful observer, he described everything he saw, including cell division in the eggs (Fig. 2–3). These develop within the uterus. The uncleaved egg has a large fluid-filled nucleus, which contains a nucleolus. Shortly before the first cell division the outline of the nucleus becomes indistinct, but by adding a little acetic acid it again becomes visible, though folded and wrinkled. Later the nucleolus disappears. All that remains is a mass of delicate, curved fibers, and these are seen only if acetic acid is added. Next thick 'strands' appear and become oriented in an equatorial plane. The granules of the egg become arranged in a regular manner. This arrangement is best seen after acetic acid treatment. The 'strands' increase in number and when the cell divides they pass into the daughter cells. You have probably identified Schneider's 'strands' with chromosomes and that is exactly

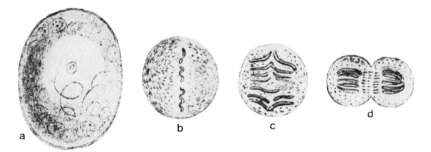

2–3 Nuclear changes during cleavage in Mesostoma embryos. *a* is an uncleaved ovum. The large clear area is the ovum itself, which contains a nucleus and nucleolus. The surrounding structures are follicle cells, which the embryo uses for food. They are not shown in *b*, *c*, and *d*. The spiral structures are sperm. *b*, *c*, and *d* show the 'strands,' which we now realize are chromosomes, and their movements during cell division. (A. Schneider, 'Untersuchungen über Plathelminthen.' *Oberhessischen Gesellschaft für Natur- und Heilkunde.* 14:69–140. 1873.)

what they were. The term 'chromosome' did not come into usage until 1888 but from now on we shall use it to avoid confusion.

If Schneider realized the importance of his observations on chromosomes in cell division, he certainly did not stress the point. The discussion in his paper is concerned with the morphology of the flatworms and the relation of these animals with other groups in the animal kingdom. It remained for others to interpret and show the importance of the phenomena that Schneider had observed.

Schneider was of the opinion that the nucleus persisted during division, though we must remember that he used acetic acid to establish this point. One could always question Schneider's interpretation, since the acetic acid treatment might have produced artifacts (abnormal structures) and the 'strands' could be so interpreted.

Bütschli also published a paper in 1873 describing cell division in a roundworm, Rhabditis. He agreed with Schneider that the nucleus persisted during cell division.

Fol, the third investigator to describe cell division in 1873, thought that the nucleus entirely disappeared during division and was re-formed in the daughter cells. This second view was shared by Flemming and Auerbach who published observations on cell division in 1874. It should be emphasized that these observers based their descriptions wholly or largely on what they observed in living eggs.

The problem of cell division was immediately recognized as being of considerable importance, and numerous investigators followed Schneider, Bütschli, and Fol. A review article on cell division and related topics was published by Professor Mark of Harvard in 1881. He quoted 194 papers (by 86 authors) which appeared in the five years from 1874 through 1878. This period was one of more or less blind experimentation and exploration. The animal and plant kingdoms were combed for favorable material. Some of the investigators observed living cells, and others worked with those that had been chemically treated. Interpretations of the observed phenomena were numerous and varied.

Some order was brought out of chaos by Flemming in 1878 (and more especially in his monograph of 1882). He was outstanding, first in selecting excellent material, namely, the epidermal cells of larval salamanders; second, in being careful to check in living cells all things that he observed in fixed and stained preparations; and third, in employing hitherto unsurpassed technical methods.

Techniques and Instruments. Before Flemming's contribution is considered in detail, we shall digress to discuss the development of techniques for preparing cells for microscopic observation. Many earlier workers used dyes in a more or less haphazard way, but in 1858 Gerlach described an adequate staining method. He found that the nuclei of preserved cells take up the dye from a dilute solution of carmine, while the rest of the cell remains unstained or becomes only slightly colored. This became a vastly improved method for observing nuclei, most of which, it must be remembered, are seen with great difficulty in the living state. Gerlach did not discover carmine; he merely perfected its use in cytology. This dye was well known to the Indians of Mexico long before the coming of the Spanish. They obtained it from the crushed and dried bodies of cochineal insects reared especially for this purpose. Later the commercial use of carmine spread to Europe. In all probability Gerlach tried it as a 'hunch.' It happened to work.

Another dye, hematoxylin, was first used successfully by Böhmer in 1865. Commercial preparations were available, derived from a tropical American tree known as logwood. This dye, like carmine, stains the nucleus.

The first synthetic aniline dye was made by Perkin. The date of this discovery is generally given as 1856, when Perkin was a lad of 18 trying to synthesize quinine. Many different aniline dyes were made later, and soon they became the principal ones used commercially. They were tried by cytologists from time to time, but it was not until the period of 1875–80 that their use was perfected. It was found that some aniline dyes, such as eosin, would stain parts of the cell not affected by carmine or hematoxylin. It was then possible to use the double-stain methods that are now standard. The nucleus could be stained deep blue with hematoxylin and the cytoplasm a pale pink with eosin. This gave a much improved picture of cell structure.

Technical advances in still another field were providing an aid to cytologists. Rapid improvement was being made in microscopes. In the 1870s Abbe, the greatest microscope designer of recent times, began his association with the Carl Zeiss optical works and increasingly fine lenses were turned out by this concern. In 1878 Abbe's oil-immersion objective was first produced (one of the initial users was the famous bacteriologist Koch). The oil-immersion lens enabled one to obtain a good image of cell structures at magnifications of more than 1,000 diameters. Technical advances in this field continued with the invention of Abbe's substage condenser, and in 1886 Zeiss produced the apochromatic objective. This is the finest lens so far developed for microscope work.

These advances meant that cytology had reached a point in its development where a person like Flemming could make a culminating advance in our understanding of cell division. He certainly did not dis-

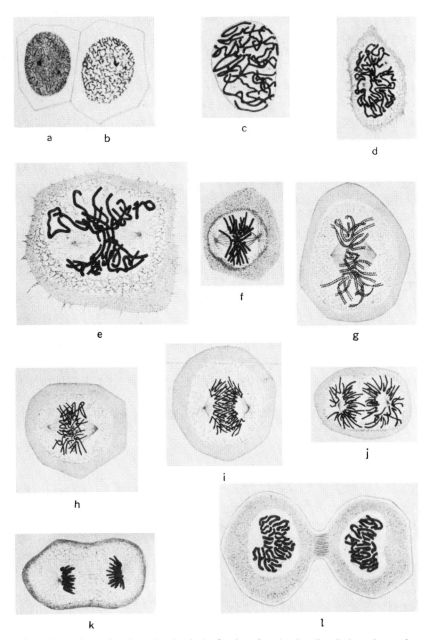

2–4 Flemming's drawing of mitosis in fixed and stained cells of the salamander embryo. The figures are arranged in sequence beginning with a resting stage in *a*. (W. Flemming, *Zellsubstanz, Kern und Zelltheilung*, 1882.) *a*. **Resting stage.** The chromosomes are invisible. The nucleus contains chromatin and two nucleoli. The nuclear membrane is present. *b*. **Late resting stage.** The chromosomes are form-

cover mitosis (neither did any single person) but we owe to him more than to any other the concept of mitosis that we hold today. After Flemming, only details were added.

Flemming's Description of Mitosis. It was well known to Flemming and his contemporaries that the structures observed in living cells might be quite different in appearance from those seen in preserved cells. In some types of living cells no nuclei could be seen, yet after staining typical nuclei were visible. In a situation of this sort the question arises 'Are nuclei present in all normal cells or can they be artifacts resulting from the treatment used in preparing the cells for study?' Flemming reasoned his answer this way. In some types of living cells, which we can call type 1, nuclei can be seen. When these cells are fixed and stained, a nucleus of characteristic shape and color appears. In other types of living cells, which we can call type 2, no nucleus can be seen. Nevertheless when type 2 cells are fixed and stained, a nucleus that in all respects is identical in appearance to the nuclei of fixed and stained cells of type 1, can be seen. Since the stained nuclei of cell types 1 and 2 have the same appearance, and the treatment is the same in both cases, the most reasonable hypothesis is that a nucleus is present, though invisible, in living cells of type 2. It seems most unlikely that cells of type 2 could be without a nucleus in the living state and that fixation and staining could produce an artifact that was identical to the nuclei of fixed and stained type 1 cells. Flemming made an attempt to apply this type of reasoning to all cell structures, and in every case he tried to use the living cell as the basis of reference. Structures that could never be seen in living cells and that made their appearance only after fixation and staining must be regarded as questionable.

The Resting Stage. Flemming's studies led to this concept of mitosis (Figs. 2–4 and 2–5). A resting stage cell is one not in mitosis. The nucleus is spherical and generally occupies the central region of the cell. A nuclear membrane is present. In living resting stage cells the nucleus does not seem to have any internal structure. After fixation

---

ing. The nucleoli are disappearing. *c.* **Prophase.** The chromosomes have formed. No nucleoli. The nuclear membrane is still present. (The cytoplasm is not shown.) *d.* **Metaphase.** The nuclear membrane has disappeared. Two centrioles, each with a tiny aster, are shown. *e.* **Metaphase.** The centrioles and the astral rays surrounding them are distinct. The chromosomes are moving to the middle of the cell. *f.* **Metaphase.** The spindle has formed between the two centrioles. *g.* **Metaphase.** This is an exceptionally good preparation. Each chromosome is seen to be double, that is, each is composed of two chromatids. *h–j.* **Anaphase.** The chromosomes are moving apart and one group is approaching each centriole. *k.* **Telophase.** The chromosomes have separated into two groups. The spindle has nearly disappeared and the astral rays are becoming indistinct. *l.* **The cell is dividing.** The chromosomes are being surrounded by a nuclear membrane and shortly each will be in the resting stage.

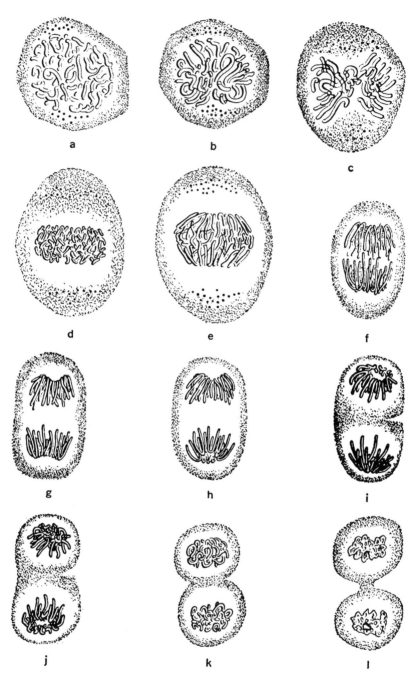

a   b   c

d   e   f

g   h   i

j   k   l

2–5 Flemming's drawings of mitosis in living epidermal cells of a salamander larva. The drawings are arranged in sequence, beginning with a prophase in *a* and ending with the two daughter nuclei in *l*. The nuclear membrane, asters,

and staining, an irregular network of strands and granules (chromatin) can be detected. In addition, one or more large spherical granules, the nucleoli, are present. Chromosomes cannot be seen in either the living or the preserved resting stage nucleus.

Prophase. Changes in the nucleus are the first indications that mitosis is under way. Long, delicate threads, the chromosomes, make their appearance. At first they are not easy to see, but with the passage of time they become increasingly distinct. Mitosis is a continuous process, but for descriptive purposes we divide it into a number of stages. When chromosomes first become visible we say that the prophase stage has begun. If prophase chromosomes are examined carefully, they are seen to be double structures, each chromosome being composed of two long strands, the *chromatids,* lying side by side. It should be emphasized that only in the very best preparations is it possible to see the chromatids. In most instances, and this is true today, only the entire chromosome is seen. Flemming was able to see the duplicate prophase chromosomes both in living and preserved salamander cells. During prophase the nucleoli become smaller and smaller and eventually they disappear.

Metaphase. Prophase ends and the next stage, metaphase, begins with the disappearance of the nuclear membrane. By this time the chromosomes have become very distinct. In stained preparations they are prominent cell structures, grouped together in the center of the cell. Early in metaphase the spindle and asters become prominent. The spindle is given this name because of its shape, which might be compared to that of a chicken's egg that is pointed at both ends. In the living cell the spindle appears as a transparent body. In fixed and stained cells there are one or more tiny granules, the *centrioles,* at each end. One can also see long strands, the *spindle fibers,* connecting the two centriole regions. At metaphase the chromosomes become arranged in a plate perpendicular to the long axis of the spindle. The *asters* are observed in fixed and stained cells as a series of fibers radiating out from the centrioles.

Anaphase. Metaphase ends and the next stage, anaphase, begins with the separation of the chromosomes into two groups. One group goes to each pole of the spindle. Flemming thought it possible that the double nature of the prophase chromosomes might be of significance in this respect. Could it be that each chromosome duplicates itself, forming two chromatids, and that at anaphase one chromatid goes to one pole

spindle, and centrioles are not shown (W. Flemming, *Zellsubstanz, Kern und Zelltheilung,* 1882).

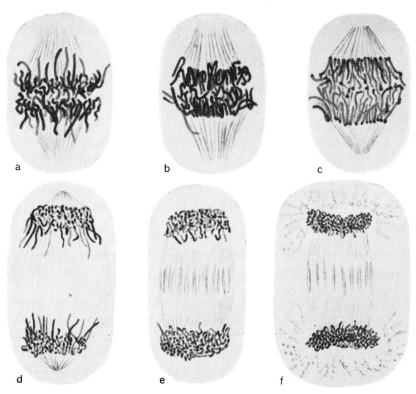

2-6 Mitosis in the lily (W. Flemming, *Zellsubstanz, Kern und Zelltheilung,* 1882).

and the other chromatid to the opposite pole of the spindle? (Flemming's belief was found to be true five years later by van Beneden.)

Telophase. The two groups of chromosomes move to the poles of the spindle. When they arrive there the last stage in mitosis, telophase, begins. The chromosomes become increasingly less distinct and the nuclear membrane is re-formed. The spindle and asters begin to disappear. The cell as a whole now divides into two daughter cells. The plane of division cuts across the spindle at the equator. As a result, each daughter cell contains a group of chromosomes. Eventually it becomes impossible to see the chromosomes; the cell has entered the resting stage once more. It should be emphasized that the term 'resting' means that the nucleus is not in mitosis. It does not signify a lack of metabolic activity.

These nuclear changes, known as mitosis, were observed in so many different kinds of animal cells that Flemming believed that they must be a universal feature of living organisms. The nuclei of plants were found to behave in an almost identical manner. Figure 2–6 shows mitosis in a lily. The chromosome stages are identical with those in the salamander but the lily, like most plants, differs from animals in lacking centrioles and asters.

Our general conclusions based on the work of cytologists up to 1882 are these: cells come from pre-existing cells, nuclei from pre-existing nuclei, and choromosomes from pre-existing chromosomes.

After studying Flemming's illustrations of chromosomes, would you have thought that all of the chromosomes of a cell are more or less alike, or that each was different from every other one in the chromosome set? This is a question that will be of the greatest importance a little later in the story.

A comment on scientific method might be inserted at this point. Flemming's use of living material had the great advantage of allowing him to work out the sequence of stages in mitosis. It was possible to establish that the events in mitosis begin with prophase, pass through metaphase, anaphase, and end with telophase. If he had studied only fixed and stained material it would have been difficult to establish any such relationship. Put yourself in his position. Would it be necessary to assume any relation between a nucleus in the resting stage and one in metaphase? If you did assume a relation could you prove it from a study of fixed and stained cells?

SUGGESTED READINGS

Baker, J. R. 1948–55. 'The cell theory: a restatement, history and critique.' *Quarterly Journal of Microscope Science.* 89:103–25; 90:87–108, 331; 93: 157–190; 94:407–40; 96:449–81.

Conn, H. J. 1928–33. 'History of staining.' *Stain Technology.* 3:1–11, 110–21; 4:37–48; 5:3–12, 39–48; 7:81–90; 8:4–18.

De Robertis, E. D. P., W. W. Nowinski, and F. A. Saez. 1960. *General Cytology.* Saunders. Chapter 1 has a brief history of cytology.

Hertwig, O. 1895. *The Cell.* Macmillan. An English translation from the German original.

Mark, E. L. 1881. 'Maturation, fecundation, and segmentation of Limax campestris Binney.' *Bulletin Museum Comparative Zoology.* 6:173–625.

Munoz, F. and H. A. Charipper. 1943. *The Microscope and Its Use.* Chemical Publishing Co. Chapter 1 discusses the early development of the microscope.

Sharp, L. W. 1934. *An Introduction to Cytology.* McGraw-Hill. Chapter 26, 'Historical Sketch.'

Wilson, E. B. 1928. *The Cell in Development and Heredity.* Macmillan.

Fertilization and Gamete Formation

At the time when some cytologists were studying the chromosomal events during mitosis, others were investigating fertilization and the formation of ova and sperm. These studies were to form the basis of our understanding of heredity, which was to come in the early years of the twentieth century.

FERTILIZATION

The elementary fact of fertilization, namely, that a sperm is required to initiate development of the ovum, was discovered by Prevost and Dumas in 1824. At this time the precise role of the sperm was not understood. In 1854, Newport observed in the frog that the sperm actually penetrates the ovum. A full understanding of this event had to wait until it was realized that both the ovum and the sperm are cells. Schwann's belief that the ovum was a cell was not shared by many cytologists, but the work of Gegenbauer in 1861 seemed to convince most workers that this was so. Several years later it was also established that the sperm was a single cell. Inheritance, then, was based on the transmission of cells—an ovum from the mother and a sperm from the father.

Fertilization in the Sea Urchin. In 1873–4 several investigators reported that two nuclei could be seen in the ovum soon after fertilization and before cell division had begun. It remained for Hertwig (1875) to demonstrate for the sea urchin that one of these nuclei was the nucleus of the ovum and the other was derived from the sperm. He found that these two nuclei approached each other, made contact, and in a slightly later stage only one nucleus was present (Fig. 3–1). In

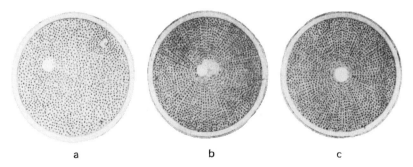

a   b   c

3-1 Hertwig's figures of sea-urchin embryos showing the nuclear events in fertilization. *a* shows an embryo 5 minutes after ova and sperm were mixed. The egg nucleus is the clear area on the left side of the embryo. The sperm nucleus is in the upper right portion of the embryo. *b* is an embryo 10 minutes after the ova and sperm were mixed. The two nuclei are in contact near the center of the embryo. *c* is an embryo 15 minutes after the ova and sperm were mixed. A single nucleus is present. This is the zygote nucleus that will undergo a series of mitotic divisions to form all of the nuclei of the individual (O. Hertwig, 'Beiträge zur Kenntniss der Bildung, Befruchtung und Theilung des thierischen Eies,' *Morph. Jahrb.* 1:347–434. 1876).

Hertwig's opinion this single nucleus was the result of fusion of a maternal nucleus of the ovum and a paternal nucleus of the sperm. Almost immediately other workers came to the same conclusion. Observations were made on eggs of many different species, and it was realized that the formation of the zygote nucleus through the fusion of a *paternal pronucleus* derived from the sperm and a *maternal pronucleus* from the ovum is a general phenomenon.

Two types of material proved of the greatest usefulness in studies of fertilization: the sea urchin (a marine animal related to the starfish) and Ascaris (a parasitic worm found in the intestine of man and other mammals). The sea urchin was especially suitable because it was easy to obtain the ova and sperm, because fertilization could be carried out under the controlled conditions of the laboratory, and because of the transparency of the ova and early embryos. The adults were collected in the ocean, usually by dredging, and in the laboratory both males and females could be stimulated to shed their gametes. The ova could be collected in one dish and the sperm in another. These would number in the millions. When the two were mixed, fertilization occurred in a matter of seconds.

One of the most striking things about fertilization and early development in the sea urchin is the fact that events are synchronous in all the zygotes fertilized at one time. Thus, if one preserves embryos at successive five-minute intervals after fertilization, the sequence of nuclear events can be worked out with precision. Shortly after fertilization, the paternal pronucleus would be noticed close to the outer membrane of the ovum. At later times it would be found progressively closer to the maternal pronucleus, and eventually fused with it.

Fertilization in Ascaris. As cytological material the sea urchin has one serious defect: its chromosomes are small and numerous. One can observe the general events in fertilization, but the details of chromosome movements and changes could not be determined with ease. On the other hand, the parasitic worm Ascaris provides excellent material for studying the behavior of chromosomes since it has only four chromosomes and these are large and stain successfully. As a consequence the detailed nuclear events in fertilization were first observed in Ascaris.

The process of fertilization in Ascaris was described by van Beneden in 1883 and by others such as Boveri in 1888. Boveri's figures, as reproduced in Fig. 3–2, will be the basis of our account of fertilization. (For the present the reader should ignore the legend for this figure, since it cannot be fully understood until the entire chapter has been read.) The first figure, *a*, shows a section of the entire ovum shortly after fertilization. The paternal pronucleus is in the lower right-hand quadrant. It contains two chromosomes. The structure forming a wrinkled cap immediately above it is the acrosome, which is the portion of the sperm head composed of Golgi material. In the center of the ovum there is a dark granular area. This is the centrosome, which was formed by a part of the sperm lying immediately behind the sperm nucleus. There are two structures near the top of the figure. The one within the ovum is the maternal pronucleus. It contains two chromosomes. The other structure, which is attached to the top of the ovum is a polar body. It can be seen in *b*, *c*, and *e* as well. For the present we shall disregard it since it is concerned with meiosis—a subject to be considered in the last part of the chapter. In *b* the maternal and paternal pronuclei have moved somewhat closer and their chromosomes have become indistinct. In *c* the chromosomes in both pronuclei have become elongated and coiled. Two centrioles have appeared in the centrosome material. In *d* the centrosome itself has divided, half being centered around each centriole. The two centrioles, with their associated centrosomes, move farther apart in *e*. In *f* they are on opposite sides of the cell with a spindle between them and an aster radiating out from each. During this period considerable changes have been occurring in the pronuclei. In *d* the chromosomes have shortened and it can be seen that each pronucleus contains two. A further shortening of the chromosomes is apparent in *e*. During the interval between *e* and *f* the membranes around both the maternal and paternal pronuclei disappear and in *f* the four choromosomes have entered the spindle. The mitotic stage shown in *f* is an early metaphase. Somewhat later each of these four chromosomes will become double to make a total of eight, and at anaphase these will separate and four chromosomes will move to each pole of the spindle.

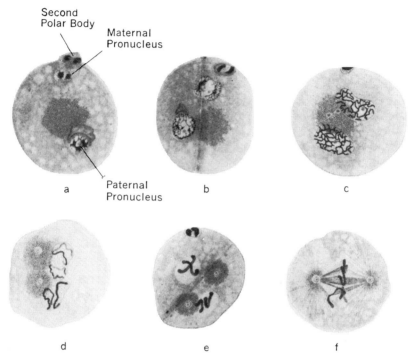

Second
Polar Body

Maternal
Pronucleus

Paternal
Pronucleus

a   b   c

d   e   f

3-2 **Fertilization in Ascaris.** In *a* the sperm has entered the ovum and formed the paternal pronucleus. The maternal chromosomes have undergone the second meiotic division. This resulted in a maternal pronucleus with 2 chromosomes and a second polar body with 2 chromosomes. The dark structure in the center of the egg is the centrosome brought in by the sperm. In *b* the maternal and paternal pronuclei have enlarged and are approaching each other. In *c* the pronuclei are continuing to enlarge. Notice the two granules between the pronuclei and in the centrosome substance. These are the centrioles. In *d* it can be seen that each pronucleus has 2 chromosomes. The centrioles are moving apart and the surrounding centrosome substance is dividing into two portions. In *e* the centrioles and the associated centrosomes have nearly reached opposite sides of the egg. The 2 chromosomes of each pronucleus are more prominent than before. The second polar body is still attached to the top of the egg. In *f* the first mitotic division of the embryo has begun. Four chromosomes, the diploid number for this species, can be seen on the spindle. These were derived from the 2 pronuclei. At this division each of these 4 chromosomes will split and the two cells that result from the division will each receive 4 chromosomes (Th. Boveri, 'Die Befruchtung und Teilung des Eies von Ascaris megalocephala,' *Jenaische Zeit.* 22:685–882. 1888).

The chromosome number of the zygote, therefore, is four. Half of this total is provided by the paternal pronucleus and half by the maternal pronucleus. The number of chromosomes in a pronucleus is spoken of as the *haploid* number and the number in the zygote is the *diploid* number. It was clear from the work of van Beneden, Boveri, and others that the chromosomal contribution of each parent to the zygote is equal. So far as one could tell the chromosomes in the maternal pronucleus were morphologically equivalent to those in the paternal pronucleus.

Further study revealed that throughout the animal kingdom similar events are observed with only a few exceptions. Fertilization involves the combination of a haploid pronucleus derived from the sperm and a haploid pronucleus derived from the ovum. Their pooled chromosomes form the diploid number of the zygote. Since the increase in cell number during embryonic development is through mitosis, all the cells of the embryo and adult should be expected to contain the diploid number of chromosomes. Research has shown this to be true with only a few exceptions.

THE FORMATION OF GAMETES

An important problem was raised by these discoveries of the chromosomal events during fertilization: if the nuclei of embryonic and adult cells are diploid, how do the nuclei of ova and sperm become haploid? Ascaris provided excellent material for the study of this problem and the observations of van Beneden, Boveri, and Hertwig established the essential points during the 1880s, first solving the problem in the ovum and later in the sperm. They discovered that there are two unusual cell divisions during the formation of gametes. As a result of these divisions diploid cells have their chromosome numbers reduced to the haploid condition. These two divisions are highly modified mitotic divisions; they are known as the *meiotic divisions*. The process itself is *meiosis*. The relation between mitosis and meiosis can be brought out by a description of the chromosomal changes during the formation of the ovary and of mature ova.

Mitosis in the Early Ovarian Cells. The ovary of Ascaris begins to form early in development. At first it consists of a few cells and in the course of time these divide to form the tremendous number comprising the ovary of the adult. *This increase in the number of cells is brought about by mitosis.* In mitosis each chromosome duplicates itself at every cell division so the number of chromosomes remains constant from one cell generation to the next. So far as individual cells are concerned this is what occurs: the Ascaris nucleus contains four chromosomes as the diploid number; before every cell division there is a duplication of each of these four chromosomes to give a total of eight chromatids; at anaphase the chromatids are separated, four going to each daughter cell. This process is repeated with the result that all the cells of the ovary are diploid.

Meiosis in the Female. Many of these ovarian cells become enlarged and form ova. The nuclei of these are diploid. The ovum of Ascaris remains diploid until it has been released from the ovary and entered by a sperm. The ovum nucleus then undergoes a series of two meiotic divisions that leads to each of the resulting cells having the haploid

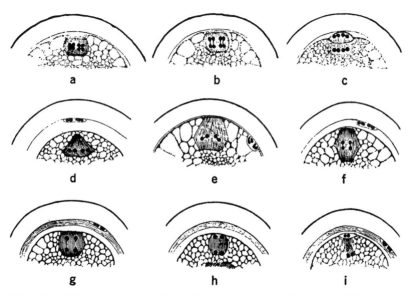

3-3 **Meiosis in Ascaris eggs.** *a* shows the upper portion of the egg and its nucleus. Previously the 4 chromosomes have undergone synapsis to form two pairs. Each chromosome then duplicated itself. The result is 2 tetrads, each composed of 4 chromatids. In this figure the tetrads are in the metaphase of the first meiotic division. *b* is the anaphase of the first meiotic division. Each tetrad has divided into 2 dyads. *c* the first meiotic division is complete. The first polar body has pinched off from the egg. It contains 2 dyads. The egg likewise contains two dyads. *d* the second meiotic division has begun and the 2 dyads are in the spindle. The first polar body with its chromosomes is beneath one of the egg membranes. It can be seen in all of the remaining figures except *h*. In *e* the dyads are rotating prior to their separation. *f* is a metaphase of the second meiotic division. *g* is the anaphase of the second meiotic division. *h* is the telophase of the second meiotic division. *i* the second meiotic division is complete. The second polar body has formed and it contains 2 chromosomes. The egg nucleus also contains 2 chromosomes (Th. Boveri, 'Die Bildung der Richtungskörper bei Ascaris megalocephala und Ascaris lumbricoides,' *Jenaische Zeit.* 21:423-515. 1887).

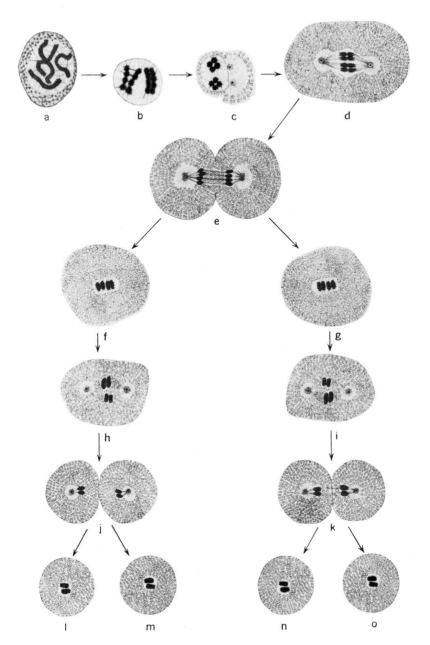

3–4   Meiosis in Ascaris males. The diploid chromosome number in Ascaris is 4. The cells of the testis that will later form the sperm are diploid as shown in *a*. *b* shows a nucleus near the beginning of meiosis. The 4 chromosomes are undergoing synapsis. As meiosis continues each chromosome becomes shortened until it forms a tiny sphere. During this process each chromosome splits. As a result each of the 2 pairs of synapsed chromosomes forms a tetrad, as shown in *c*. At the first

number of chromosomes. The process of meiosis in the Ascaris ovum is shown in Fig. B3–3, which is reproduced from the work of Boveri.

The First Meiotic Division of the Ova. At the onset of meiosis each of the four long chromosomes (as shown in Fig. 3–2*f*) becomes condensed to form a tiny sphere. Next the chromosomes come together in pairs, a process that is known as *synapsis.* The chromosomes do not fuse during synapsis, they merely come close to one another. The next event that occurs is that each chromosome becomes duplicate. Thus, each of the two pairs of chromosomes becomes a group of four, such a group being known as a *tetrad.* The first of Boveri's figures, namely 3–3*a,* shows an ovum in this condition, which is the metaphase of the first meiotic division. In it we see the chromosomes grouped into two tetrads. In *b* the tetrads are being divided and in *c* they have separated completely. Half of each tetrad, or a *dyad,* goes to each pole of the spindle. It will be noticed that the spindle is not in the center of the cell but instead it is at the periphery. Inasmuch as the cell will divide across the equator of the spindle, the result will be two cells of very unequal sizes. The large cell resulting from the division is the ovum and the small cell is the *first polar body.* The chromosomes that enter the first polar body are morphologically and numerically equivalent to those that remain in the ovum.

The Second Meiotic Division of the Ova. In *d* the first polar body is well separated from the ovum and the two dyads within the ovum are on the spindle of the second meiotic division. *At this division the chromosomes do not duplicate themselves.* Consequently the dyads are divided and as a result two chromosomes go to each pole of the spindle. This second meiotic division divides the cell unequally, as did the first, the result being a large ovum and a tiny second polar body. At the end of the second, and last, meiotic division there are only two chromosomes in the Ascaris ovum. The nuclear membrane forms around these two chromosomes, the haploid number, and in this manner the maternal pronucleus is produced.

The subsequent history of the maternal pronucleus has been discussed as an aspect of fertilization and Fig. 3–2 should be re-studied (the maternal pronucleus in 3–3*i* is in the same stage as in 3-2*a*).

---

meiotic division the 2 tetrads enter the spindle (*d*) and are divided, half of each tetrad (a dyad) going to each pole as shown in *e.* As a result of the first meiotic division 2 cells are formed (*f, g*). Each of these contains 2 dyads. In the second meiotic division (*h, i, j, k*) the dyads of the 2 cells are pulled apart. At the end of this division there are 4 cells (*l, m, n, o*). Each of these contains 2 chromosomes, the haploid number. There is no further division of these 4 cells and they develop directly into sperm (A. Brauer, 'Zur Kenntniss der Spermatogenese von Ascaris megalocephala,' *Arch. Mikr. Anat.* 42:153–213. 1893).

OUTLINE OF MEIOSIS AND

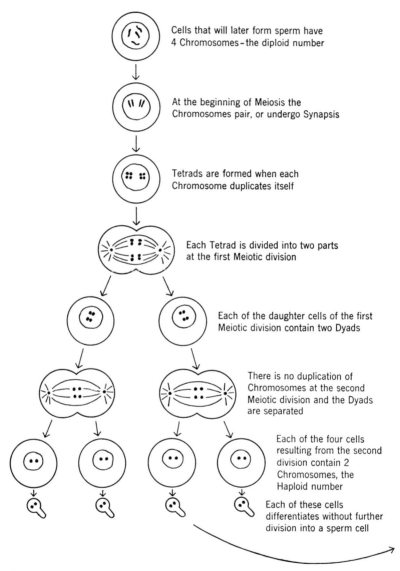

Cells that will later form sperm have 4 Chromosomes – the diploid number

At the beginning of Meiosis the Chromosomes pair, or undergo Synapsis

Tetrads are formed when each Chromosome duplicates itself

Each Tetrad is divided into two parts at the first Meiotic division

Each of the daughter cells of the first Meiotic division contain two Dyads

There is no duplication of Chromosomes at the second Meiotic division and the Dyads are separated

Each of the four cells resulting from the second division contain 2 Chromosomes, the Haploid number

Each of these cells differentiates without further division into a sperm cell

3–5

FERTILIZATION IN ASCARIS

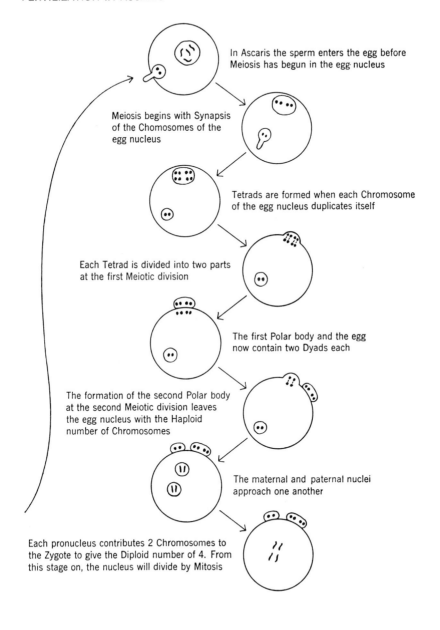

In Ascaris the sperm enters the egg before Meiosis has begun in the egg nucleus

Meiosis begins with Synapsis of the Chomosomes of the egg nucleus

Tetrads are formed when each Chromosome of the egg nucleus duplicates itself

Each Tetrad is divided into two parts at the first Meiotic division

The first Polar body and the egg now contain two Dyads each

The formation of the second Polar body at the second Meiotic division leaves the egg nucleus with the Haploid number of Chromosomes

The maternal and paternal nuclei approach one another

Each pronucleus contributes 2 Chromosomes to the Zygote to give the Diploid number of 4. From this stage on, the nucleus will divide by Mitosis

Meiosis in the Male. The observation that the paternal pronucleus was haploid, yet the male diploid in its body cells suggested that a process similar to that just described must also occur in the male. A study of sperm formation in Ascaris showed this to be the case (Fig. 3–4). The last two cell divisions before a sperm forms are meiotic divisions. As in the egg, the four chromosomes form two pairs and each chromosome duplicates itself. The result is two tetrads each composed of four chromatids. During the first meiotic division the tetrads are divided and half of each goes into each of the daughter cells. Not only is nuclear division equal but cell division is also equal, which is in contrast to the situation in the ova. At the next division the dyads are divided between the two daughter cells, which are again of equal size. Thus, from one cell with four chromosomes, and by means of two meiotic divisions, we end with four cells each with two chromosomes. Each of these four haploid cells then develops without further division into a sperm cell.

The essential difference between meiosis and mitosis is this: in mitosis there is one duplication of every chromosome for each cell division; in meiosis there is only one duplication of every chromosome for the two meiotic divisions. As a consequence, in mitosis the chromosome number remains constant from one cell generation to the next; in meiosis the two meiotic divisions form cells with the haploid number of chromosomes.

With full realization that the nuclear events associated with maturation and fertilization were important biological phenomena, cytologists examined many species of animals and plants. It was found that the reduction divisions leading to haploid pronuclei occur throughout the animal and plant kingdoms. In short, another principle of almost universal application (a few exceptions were found) had been discovered. The facts as outlined in this section were generally, though not universally, believed by 1890.

A summary of meiosis and fertilization in Ascaris is given in Fig. 3–5.

SUGGESTED READINGS

Hertwig, O. 1876. 'Beiträge zur Kenntniss der Bildung, Befruchtung und Theilung des theirischen Eies.' *Morphologisches Jahrbuch* 1:347–435. Figure B3–1 in this chapter is taken from this article of Hertwig.

The references to Mark, Sharp, and Wilson given at the end of Chapter 2 will serve for the present chapter as well.

The Nucleus and Heredity

The middle years of the 1880s witnessed several attempts to see if inheritance was controlled by some definite part of the cell. We might have expected this to be the case when we realize that cytologists, in a decade of unparalleled discovery, had worked out the essentials of mitosis, fertilization, and meiosis.

Haeckel's Hypothesis of the Nuclear Control of Inheritance. An effort to find a cytological basis for inheritance was made as early as 1866 by Haeckel, who postulated that the nucleus was responsible for the transmission of the inherited features of an organism. The data available to Haeckel in 1866 were not sufficient to test this hypothesis. As E. B. Wilson, the great American cytologist, was to remark some years later, it was a lucky guess. If a lucky guess of this sort had been made by some obscure scientist, it is probable that its influence on subsequent events would have been negligible. But Haeckel was a leader in the field of biology in his day. An idea of his, no matter how slight the factual basis, would have been noticed. It is conceivable, therefore, that Haeckel's hypothesis of nuclear control of inheritance helped to prepare others for thinking and experimenting along these lines.

Nägeli's Idioplasm Theory. In 1884 Nägeli suggested that a substance which he called the *idioplasm* was responsible for inheritance. The idioplasm was thought to be an invisible chemical network that extended throughout the cell and from cell to cell. Nägeli did not observe the idioplasm in cells. He invented it to account for inheritance. He did not regard it as a highly stable material, but as one that might change during development, or as the result of nutrition or other external conditions. In any event it must return to the original condition in the embryo. Nägeli did considerable theorizing on the subject of inheritance, but his concept of possible mechanisms was extremely vague. His hypothesis was nearly impossible to test, and hence it could be of no real usefulness in directing efforts to profitable experimentation.

Early Evidence for the Nuclear Control of Inheritance. In 1884–5 four German scientists, working independently, came to the conclusion that the physical basis of inheritance must lie in the chromosomes. They were Hertwig, Strasburger, Kölliker, and Weismann. The first three were primarily laboratory scientists. For at least a decade they had been leaders in the analysis of problems concerned with the nucleus. Weismann, on the contrary, is remembered largely for his theoretical work.

These four men believed that the chromosomes were the physical basis of inheritance for the following reasons.

1. Even though inheritance was not well understood, it seemed that both parents have an equal share in transmitting their characteristics to the offspring. What is the physical basis of this equality? It was known, of course, that the only links between parent and offspring are the ovum and sperm. These two cells are about as different as any two cells could be. Usually the ovum has a mass thousands or millions of times the mass of the sperm. Ova usually contain a large quantity of cytoplasm, whereas sperm contain almost none. This would suggest that the cytoplasm was not the basis of inheritance because, if it were, it might be expected that the female's contribution would be much greater than the male's. The only parts of the sperm and ova that seemed to these four scientists to be equivalent were the nuclei. The sperm pronucleus and the egg pronucleus were identical so far as one could tell. *Perhaps this equivalence of structure was the basis of the equivalent importance of the two gametes in inheritance.* Van Beneden's description of the pronuclei in Ascaris, each with two chromosomes, seemed most suggestive.

2. During cell division, the cytoplasm and its formed structures seem to be divided passively. The chromosomes, on the other hand, go through a complicated mitosis which results in each of the daughter cells receiving exactly the same number of chromosomes. It seemed to Hertwig and the others that the significance of this complicated process might be that the nucleus was the basis of inheritance: why should the chromosomes, alone among the cell structures, be duplicated and then divided equally unless they were of great importance in inheritance?

3. The complex chromosomal changes during meiosis were understandable in terms of keeping the chromosomes constant from generation to generation. There was no similar phenomenon for any other

cell structure. Since inheritance was an intergeneration phenomenon and the chromosomes seemed to be the only cell structures that were transmitted in an exact way from one generation to another, perhaps the chromosomes were of importance in inheritance.

4. Finally, there was a more direct test of nuclear function in regenerating protozoa. The forms selected for this work were single-celled organisms with one nucleus. It was possible to cut the animals into two parts, one part containing cytoplasm and the other cytoplasm and the nucleus. Both parts healed. The part without a nucleus lived for some time, but it was unable to regenerate to form a whole animal, and it was incapable of reproduction. The part with the nucleus could regenerate a whole animal and could reproduce normally.

These observations were suggestive, but they did not 'prove' that the nucleus was the physical basis of inheritance. The fact that chromosomes appeared to be the only cell structure that remained constant from cell to cell, and from generation to generation, *could* mean that inheritance was by way of the chromosomes. Many famous cytologists believed that a good working hypothesis was 'The nucleus is important in heredity.'

In the next chapter, we shall learn that in the year 1900 the rediscovery of a scientific paper written much earlier by Mendel put the subject of inheritance in an entirely new light. It is of interest, therefore, to summarize the advances that those cytologists interested in heredity had made up to the year Mendel's results became generally known. Such a summary was given retrospectively by Wilson in 1914:

The work of cytology in its period of foundation laid a broad and substantial basis for our more general conceptions of heredity and its physical substratum. It demonstrated the basic fact that heredity is a consequence of the genetic continuity of cells by division, and that the germ-cells are the vehicle of transmission from one generation to another. It accumulated strong evidence that the cell-nucleus plays an important role in heredity. It made known the significant fact that in all the ordinary forms of cell-division the nucleus does not divide *en masse* but first resolves itself into a definite number of chromosomes; that these bodies, originally formed as long threads, split lengthwise so as to effect a meristic division of the entire nuclear substance. It proved that fertilization of the egg everywhere involves the union or close association of two nuclei, one of maternal and one of paternal origin. It established the fact, sometimes designated as 'Van Beneden's law' in honor of its discoverer, that these primary germ-nuclei give rise to similar groups of chromosomes, each containing half the number found in the body-cells. It demonstrated that when new germ-cells are formed each again receives only half the number characteristic of the body-cells. It steadily accumulated evidence, especially through the admirable studies of Boveri, that the chromosomes of successive generations of cells, though commonly lost to view in the resting nucleus, do not really lose their individuality, or that in some less obvious way they conform to the principle of genetic continuity. From these facts followed the far-reaching conclusion that the nuclei of the body-cells are diploid or duplex structures, descended equally from the original maternal and paternal chromosome-groups of the fertilized egg. Continually receiving confirmation by the labours of later years, this result gradually took a central place in cytology; and about it all more specific discoveries relating to the chromosomes naturally group themselves.

All this had been made known at a time when the experimental study of heredity was not yet sufficiently advanced for a full appreciation of its significance; but some very interesting theoretical suggestions had been offered by Roux, Weismann, de Vries, and other writers. While most of these hardly admitted of actual verification, two nevertheless proved to be of especial importance to later research. One was the pregnant suggestion of Roux (1883), that the formation of chromosomes from long threads brings about an alignment in linear series of different materials or 'qualities.' By longitudinal splitting of the threads all the 'qualities' are equally divided, or otherwise definitely distributed, between the daughter-nuclei. The other was Weismann's far-seeing prediction of the reduction division, that is to say, of a form of division involving the separation of undivided whole chromosomes instead of the division-products of single chromosomes. This fruitful suggestion (1887) pointed out a way that was destined to lead years afterwards to the probable explanation of Mendel's law of heredity.

Such, in bird's-eye view, were the most essential conclusions of our science down to the close of the nineteenth century.

SUGGESTED READINGS

Roberts, H. F. 1929. *Plant Hybridization before Mendel.* Princeton University Press.

Wilson, E. B. 1900. *The Cell in Development and Inheritance.* Macmillan. This, the second edition of the classic in cytology, gives a fascinating description of the state of the science in 1900. The Introduction and Chapter 9 are of special interest.

Wilson, E. B. 1914. 'Croonian Lecture: The bearing of cytological research on heredity.' *Proceedings of the Royal Society.* B. *88*:333–52.

Zirkle, C. 1951. 'The knowledge of heredity before 1900' in *Genetics in the 20th Century,* edited by L. C. Dunn. Macmillan.

5

Mendel—1900

During the entire period from Darwin's attempted synthesis of the facts of inheritance down to 1900, a scientific paper that was to revolutionize our understanding of heredity lay unappreciated on the shelves of many libraries. The article itself had been published in 1866. In it the author, Gregor Mendel, presented some of the results of his experiments in crossing varieties of garden peas.

The Discovery of Mendel's Paper. The 'discovery' and appreciation of the importance of Mendel's paper is a very dramatic incident in the history of science. Three individuals, de Vries, Correns, and Tschermak, independently in the year 1900, realized the great importance of Mendel's work.

During the 1890s there was renewed interest in plant hybridization. The three scientists who 'discovered' Mendel—de Vries, Correns, and Tschermak—were doing breeding experiments of their own. Each of them independently came to more or less the same conclusions that Mendel had expressed in 1866 *before they knew of Mendel's paper.* This is another example of a frequent happening in science. When the field is 'ready,' the discovery is certain to be made. If Mendel had never lived, the history of genetics would not have been greatly different. About the year 1900, either he would be rediscovered or, had he never lived, others would reach essentially the same conclusions as he had in 1866. His work was unappreciated in his own lifetime, for biologists in 1866 had neither the background nor the prescience to understand the significance of what he had accomplished.

Gregor Mendel's famous article is not a scientific paper in the usual sense, but a lecture presented to the Natural History Society of Brünn in 1865. The full results of his research were never published, but the portion that he did include, coupled with an extraordinary analysis of the data, make his paper one of the landmarks of science.

Mendel was fully aware that experiments in plant breeding had been conducted by many famous men. It was true, nevertheless, that no general principles had emerged from previous studies. To Mendel this was a serious affair, since an understanding of inheritance was essential for an understanding of evolution and he was deeply interested in Darwin's work (*The Origin of Species* appeared during the period he was conducting his experiments). He began experiments which were intended to give information on inheritance and evolution.

Peas as Experimental Material. Mendel selected peas for his experiments because they possessed many desirable features:

1. Numerous varieties of peas were available commercially. They provided the material that he studied.

2. The plants were easy to cultivate and the generation time was short.

3. The offspring of the crosses between the varieties were fertile.

4. The structure of the pea flower is such that accidental pollination was thought not to occur. The anthers that produce the pollen and the stigma where the pollen grains germinate, are completely enclosed by the petals. Normally, pollen from a flower falls on the stigma of the same flower and self-fertilization results. In those cases where crosses between varieties are desired, it is possible to remove the anthers before they mature and somewhat later, when the stigma is mature, to cover it with pollen from another flower.

Mendel's approach to the problem of inheritance was different from that of previous workers. His predecessors had concentrated on the whole organism. Usually they had crossed varieties that differed in many characters and the offspring were found to be intermediate or in rare cases more like one parent. Mendel focused his attention on specific differences and studied how these were inherited generation after generation. Some of his varieties had *round* seeds; others had *wrinkled* seeds. In all he studied seven different characters of the pea, and for each he had two varieties, as shown in the following lists:

CHARACTER AFFECTED	VARIETIES
Seed shape	*round* or *wrinkled*
Seed color	*yellow* or *green*
Seed coat color	*colored* or *white*
Pod shape	*inflated* or *wrinkled*
Pod color	*green* or *yellow*
Flower position	*axial* or *terminal*
Stem length	*long* or *short*

Crosses of Plants with Contrasting Characters. First, he made sure that all of his varieties would breed true. Once this was established he made crosses between all of the pairs just listed. The results were most unexpected in the light of earlier experiments by other plant breeders. The offspring were never intermediate but were always like one of the parents. When peas with *round* seeds were crossed with peas with *wrinkled* seeds, for example, the offspring were plants with *round* seeds. Mendel spoke of the form that appeared in the offspring as *dominant* in comparison to the form that did not appear, which he called *recessive.* (Dominance can be determined only by making a cross and observing the type of offspring obtained; it could not be predicted before the experiment was performed merely by examining the parent plants.) The varieties that Mendel used were found to have these relationships:

DOMINANT	RECESSIVE
round seed	*wrinkled* seed
yellow seed	*green* seed
colored seed coat	*white* seed coat
inflated pod	*wrinkled* pod
green pod	*yellow* pod
axial flowers	*terminal* flowers
long stem	*short* stem

Over the course of years, geneticists have introduced some terms that make it easier to discuss crosses. The original parental generation is abbreviated P. The offspring of the P generation is the first filial or F_1 generation. The offspring of the F_1 is the second filial generation or F_2, the third is the F_3, and so on. It is also customary to describe crosses in terms of the character. Thus, the cross of a plant with round seeds with a plant with wrinkled seeds is shortened to *round* $\times$ *wrinkled.*

The F_2 Generation. Some plant breeders might have stopped the experiments after a single cross had determined dominant and recessive characteristics. The results, after all, were uniform and clear cut. The F_1 plants were always like one of the parents. Mendel, however, continued his crosses and was careful to realize that although an F_1 of the *round* $\times$ *wrinkled* might be identical with the *round* parent, its parentage was different. Perhaps its genetic behavior would reflect the different origin.

Since peas are self-fertilizing, the F_1 plants pollinated their own ovules and gave the F_2, and when Mendel studied the F_2 plants he found that both dominant and recessive characters were present. Now he did a simple though revolutionary thing: he counted the number of individuals of each type. In every cross there was a ratio of 3 dominant to 1 recessive. His results can be summarized as follows:

P	F_1	F_2 COUNTS	RATIO
round × wrinkled	round	5,474 round	
		1,850 wrinkled	2.96:1
yellow × green	yellow	6,022 yellow	
		2,001 green	3.01:1
colored × white	colored	705 colored	
		224 white	3.15:1
inflated × wrinkled	inflated	882 inflated	
		299 wrinkled	2.95:1
green pods × yellow pods	green	428 green	
		152 yellow	2.82:1
axial × terminal	axial	651 axial	
		207 terminal	3.14:1
long × short	long	787 long	
		277 short	2.84:1

These results would suggest that the rules of inheritance were the same, irrespective of the varieties being crossed. The F_1 plants were always of one type, which resembled one of the parents. In the F_2, two classes appeared and the frequency was 75 per cent dominants and 25 per cent recessives, or a ratio of 3:1.

The F_3 Generation. Mendel continued his experiments and obtained an F_3 generation. We may take as an example of his work the *round* × *wrinkled,* which in the F_2 gave 75 per cent *round* and 25 per cent *wrinkled.* He allowed a number of the *wrinkled* plants to self-fertilize to give an F_3, and found that all bred true, that is, only *wrinkled* plants were obtained in the F_3. The F_2 *round* plants gave two results:

1. One-third (193 of 565 plants) bred true, giving *round* plants in the F_3.
2. Two-thirds (372 of 565 plants) gave *round* and *wrinkled* in a ratio of 3:1. (On the basis of their genetic behavior these were like the F_1 plants.)

Mendel's Hypothesis. Mendel explained these results in this way: Let us assume that the *round* variety is round because it has a *gene* **R**, and the *wrinkled* variety is wrinkled because it has gene **r**. (Mendel did not use the term gene but spoke of 'factors' or 'traits.' It will be simpler for us to use the modern term gene from the very beginning and note the gradual change in its meaning. For the present we shall understand it to be the basis of an inherited character.) If a cross is made between the *round* and *wrinkled* varieties, the gametes of the *round* plant will have **R** and the gametes of the *wrinkled* plant will have **r**. Fertilization will produce an F_1 with both genes, **Rr**. In appearance this plant is *round,* **R** being dominant and **r** recessive. We speak of the appearance of the individual as its *phenotype* and the genetic composition of the individual as its *genotype*. Thus, the phenotype of this F_1 plant is *round* and its genotype is **Rr**. Plants of the **Rr** type will produce gametes and Mendel assumed that any single gamete would contain either **R** or **r** *but never both*. He also assumed that gametes containing **R** and gametes containing **r** are produced in equal numbers. If the union of F_1 gametes is at random, then we will obtain a 3:1 ratio of *round* to *wrinkled*. The entire cross could be presented in this schematic manner:

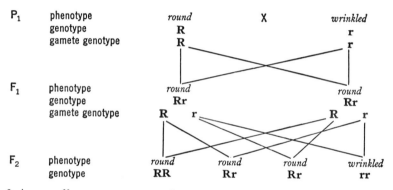

P_1	phenotype	*round*	X	*wrinkled*
	genotype	**R**		**r**
	gamete genotype	**R**		**r**
F_1	phenotype	*round*		*round*
	genotype	**Rr**		**Rr**
	gamete genotype	**R** **r**		**R** **r**
F_2	phenotype	*round* *round*	*round*	*wrinkled*
	genotype	**RR** **Rr**	**Rr**	**rr**

It is actually unnecessary to show two plants in the F_1 since both are the same. Two are used in order to make it easier to visualize the cross that gives the F_2.

This scheme provides a formal explanation of the results, namely, the origin of the 3:1 ratio. It also shows that the F_2 *round* plants are of two types. One in three of the *round* plants is pure *round*. If self-fertilized, it would breed true. The remaining ⅔ of the *round* plants are **Rr**. If these are allowed to self-fertilize there will result an F_3 ratio of 3 *round* to 1 *wrinkled*. It will be recalled that Mendel made these tests of the F_2 and the theoretical and actual results are the same.

This schematic interpretation applies to all of Mendel's crosses involving one pair of genes. Several important conclusions can be reached if the interpretation is correct:

1. Dominant and recessive genes do not affect one another. In the F_1 of the cross discussed, the genotype was **Rr**. There was no visible effect of the **r** gene, the seeds being just as round as in the pure *round* parent. When the **Rr** plant was allowed to self-fertilize both *round*

and *wrinkled* seeds were obtained. These F_2 *wrinkled* seeds were identical in appearance to the P generation *wrinkled* seeds.

2. The gametes produced by an F_1 plant of the **Rr** constitution will contain either **R** or **r**, never both.

3. The **R** and **r** types of gametes will be produced in equal numbers by an **Rr** plant.

4. Combination between gametes is a chance affair, and the frequency of different classes of offspring will depend on the frequencies of gametes. Since an F_1 plant having the **Rr** constitution will produce 50 per cent gametes of the **R** type and 50 per cent gametes of the **r** type the mathematical basis of the F_2 frequencies will be as follows:

POLLEN

	50% **R**	50% **r**
50% **R**	25% **RR**	25% **Rr**
50% **r**	25% **Rr**	25% **rr**

OVULES

Crosses Involving Two Pairs of Genes. Mendel's next step was to see if the conceptual scheme devised for crosses involving one pair of genes could be applied to crosses involving two pairs of genes. For this he used *round* and *wrinkled* as one pair and *yellow* and *green* as the other. Previous work had shown that the cross between *yellow* and *green* produced *yellow* in the F_1 and a ratio of 3 *yellow* to 1 *green* in the F_2.

When a cross was made between a plant with *round-yellow* seeds and a plant with *wrinkled-green* seeds all of the F_1 plants had *round-yellow* seeds. In the F_2 the following seed types were obtained:

315 *round-yellow*
108 *round-green*
101 *wrinkled-yellow*
32 *wrinkled-green*

One interesting thing brought out by these data is the appearance of two new seed types that were not present in either the P or the F_1 generation. These new types are *round-green* and *wrinkled-yellow*. This and the other results, however, fit perfectly into the Mendelian scheme, if we assume the complete independence in inheritance of the two pairs of genes. The cross would be as follows:

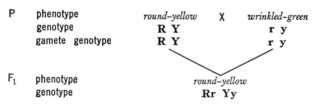

In the formation of gametes by the F_1 plant, Mendel assumed that a gamete would have only one member of a pair of genes. Thus, a gamete would have either **R** *or* **r** and in addition either **Y** *or* **y**. Four classes of gametes would be produced and these in equal frequency. The classes would be **RY**, **Ry**, **rY**, and **ry**. If an F_1 plant is allowed to self-fertilize there will be these four types of pollen and the same four types of ovules. This will give 16 possible combinations, as shown below:

POLLEN

	RY	Ry	rY	ry
RY	**RR YY** *round-yellow*	**RR Yy** *round-yellow*	**Rr YY** *round-yellow*	**Rr Yy** *round-yellow*
Ry	**RR Yy** *round-yellow*	**RR yy** *round-green*	**Rr Yy** *round-yellow*	**Rr yy** *round-green*
rY	**Rr YY** *round-yellow*	**Rr Yy** *round-yellow*	**rr YY** *wrinkled-yellow*	**rr Yy** *wrinkled-yellow*
ry	**Rr Yy** *round-yellow*	**Rr yy** *round-green*	**rr Yy** *wrinkled-yellow*	**rr yy** *wrinkled-green*

OVULES (label to the left of the table, aligned with the Ry/rY rows)

Of the 16 possible F_2 combinations, 9 will be *round-yellow,* 3 will be *round-green,* 3 will be *wrinkled-yellow,* and 1 will be *wrinkled-green.* Here is a comparison of Mendel's data with the theoretical expectation:

	ACTUAL	EXPECTED
Round-yellow	315	313
Round-green	108	104
Wrinkled-yellow	101	104
Wrinkled-green	32	35
	556	556

The figures given in the 'actual' column are those obtained by counting the seeds. The 'expected' values are computed in this manner: If we have a total of 556 plants and expect $\frac{1}{16}$ of them to be *wrinkled-green*, we find $\frac{1}{16}$ of 556, which is 35. The other expected classes are $\frac{9}{16}$, $\frac{3}{16}$, and $\frac{3}{16}$ of 556.

Testing the Hypothesis. Mendel made a further test of the adequacy of his hypothesis. On the basis of his hypothesis the F_2 plants would have four phenotypic classes and a total of nine genotypic classes (refer to the checkerboard). Genetic tests would allow him to distinguish among plants of the same phenotype but of different genotypes. His test consisted of allowing all the F_2 plants to self-fertilize to produce an F_3 and then seeing if the actual results of the crosses were the same as were expected on the basis of the hypothesis. This is what he found.

The Breeding Behavior of the F_2 Round-yellow. It can be seen from the checkerboard that $\frac{9}{16}$ of the F_2 plants are *Round-yellow*. These plants are listed in the first column of the table that follows, grouped according to genotype. These plants are all of the same phenotype but they belong to four different genotypes, namely **RRYY**, **RRYy**, **RrYY**, and **RrYy**. Although of identical appearance, these four genotypes can be distinguished on the basis of the ratios of the types of offspring they will produce following self-fertilization. These expected ratios are listed in the second column. The third column gives the number of plants that one would expect to give the ratio listed in column 2, if Mendel's hypothesis is correct. For example, Mendel expected one out of every nine *round-yellow* plants to be **RRYY**. If a plant of this genotype is self-fertilized, it would give only *round-yellow* offspring in the F_3. No other genotype will give this result. Mendel planted 315 of the F_2 *round-yellow* plants and of these 301 gave progeny. He would expect therefore $\frac{1}{9}$ of these 301 plants, or 33, to be **RRYY** and give only *round-yellow* seeds. The fourth column gives the actual results. In the example we have been using, Mendel expected that 33 of the plants would be **RRYY** and he found that 38 were of this genotype.

F_2	EXPECTED F_3 RATIOS	EXPECTED	ACTUAL
RRYY	all *round-yellow*	33	38
RRYy **RRYy**	3 *round-yellow;* 1 *round-green*	67	65
RrYY **RrYY**	3 *round-yellow;* 1 *wrinkled-yellow*	67	60
RrYy **RrYy** **RrYy** **RrYy**	9 *round-yellow;* 3 *round-green;* 3 *wrinkled-yellow;* 1 *wrinkled-green*	134	138
		301	301

The Breeding Behavior of the F_2 Round-green. Three-sixteenths of the F_2 were of this category. Mendel raised 102 of these plants and these are the results:

F_2	EXPECTED F_3 RATIOS	EXPECTED	ACTUAL
RRyy	all *round-green*	34	35
Rryy Rryy	3 *round-green; 1 wrinkled-green*	68	67
		—	—
		102	102

The Breeding Behavior of the F_2 Wrinkled-yellow. The *wrinkled-yellow* comprised $\frac{3}{16}$ of the F_2. Mendel raised 96 of these plants and these are the results:

F_2	EXPECTED F_3 RATIOS	EXPECTED	ACTUAL
rrYY	all *wrinkled-yellow*	32	28
rrYy rrYy	3 *wrinkled-yellow; 1 wrinkled-green*	64	68
		—	—
		96	96

The Breeding Behavior of the F_2 Wrinkled-green. This category comprised $\frac{1}{16}$ of the F_2. Mendel raised 30 plants of this type. The results were as follows:

F_2	EXPECTED F_3 RATIOS	EXPECTED	ACTUAL
rryy	all *wrinkled-green*	30	30

The fact that the F_2 plants gave an F_3 that did not differ materially from the expected, indicated that Mendel's conceptual scheme of inheritance was in full accord with his experimental results. In every case the actual values are surprisingly close to those expected. The expected values are based on the probability of the various types of gametes combining in a certain way. The expected and actual values are rarely identical: we should not expect them to be so any more than we should always expect five heads for every ten tosses of a coin.

Mendel went one step farther and crossed plants differing in three contrasting characters. The results were entirely according to expectation but they will not be discussed.

These are some of the conclusions that may be drawn from Mendel's experiments:

1. The most important conclusion is that inheritance appears to follow definite and rather simple rules. Mendel was able to apply the same type of explanation to the results of all of his crosses. He had

reached that stage in the development of a scientific theory where results could be predicted with a high degree of accuracy. This is one goal of a scientist.

2. When plants of two different types were crossed, there was no blending of the individual characteristics. Mendel studied seven pairs of contrasting characters. One member of each pair of contrasting characters could be thought of as dominant and the other as recessive. In a hybrid formed from crossing a pure-breeding dominant and a pure-breeding recessive, the appearance of the plant was identical with that of the dominant parent.

3. The factors responsible for the dominant and recessive condition were not modified by their occurence together in a hybrid. If two hybrids were crossed, both dominant and recessive offspring would appear. Neither the dominant nor the recessive offspring would give any evidence of contamination resulting from hybridization. In short, an F_2 recessive would be identical in genotype and phenotype to the P generation recessive.

4. When a pure-breeding plant exhibiting a dominant characteristic (**A**) is crossed with a recessive (**a**) the F_1 (**Aa**) is like the **A** plant in appearance. *Segregation* occurs in the F_2, which results in a ratio of three plants having the dominant character (one of which will be pure breeding and the other two like the F_1) to one recessive. Segregation is often called *Mendel's first law*.

5. If two pairs of genes, such as **Aa** and **Bb**, are involved in a cross, each pair acts independently so far as transmission to the next generation is concerned. This phenomenon is known as *independent assortment* and it is often spoken of as *Mendel's second law*. Its mode of operation can be understood if we consider the F_2 originating from an F_1 **AaBb** plant. So far as the phenotypes are concerned $\frac{3}{4}$ of the plants will have the **A** phenotype and $\frac{1}{4}$ will have the **a** phenotype. The same is true for the other pair of genes: $\frac{3}{4}$ of the plants will have the **B** phenotype and $\frac{1}{4}$ will have the **b** phenotype. It is entirely a matter of chance which combination of genes a given F_2 plant will receive. Thus of the $\frac{3}{4}$ that will have the **A** phenotype, $\frac{3}{4}$ will also have the **B** phenotype and $\frac{1}{4}$ will have the **b** phenotype. Of the $\frac{1}{4}$ that will have the **a** phenotype, $\frac{3}{4}$ will have the **B** phenotype and $\frac{1}{4}$ will have the **b** phenotype. So far as both characters are concerned $\frac{9}{16}$ of the F_2 ($\frac{3}{4}$ of $\frac{3}{4}$) will have both the **A** and **B** phenotypes, $\frac{3}{16}$ will have the **A** and **b** phenotypes, $\frac{3}{16}$ will have the **a** and **B** phenotypes, and $\frac{1}{16}$ will have the **a** and **b** phenotypes. The 9:3:3:1 ratio of the F_2 is due to the independent assortment of genes in the gametes of the F_1 plant.

6. The gametes will contain only one type of inherited factor of each contrasting pair. Thus the gametes of an F_1 **Aa** plant will produce gametes containing either **A** or **a**, never both. If two factors are involved, as in an **AaBb** plant the gametes will be **AB**, or **Ab**, or **aB**, or **ab**, never **Aa**, **Bb**, **ABb**, **Aab**, and so on. All possible combinations will be obtained, consisting of one member of each pair of genes. Every type of gamete will be produced in equal frequency.

It must be remembered that, without further work, these conclusions could apply only to the seven pairs of pea genes actually studied by Mendel. The fact that Mendel's rules applied to these would indicate that other genes of peas *might* behave in a similar way. The discovery that rules of inheritance could be established for peas would suggest that the same or similar rules might also apply to other plants (and possibly animals). This, of course, would have to be tested by experimentation. The great worth of Mendel's theory was that its clear and definite formulation made testing by experimentation possible. The same could not be said for any previous theory of inheritance.

Most of the important discoveries in biology turn out in retrospect to be fairly simple. Inevitably we wonder, Why did not that idea occur to someone before? Why had no one discovered these simple relations when varieties were crossed? Why had no one realized the significance of Mendel's approach during the 34 years between 1866 and 1900? These are unanswerable questions, but the following facts are germane to the last one. Mendel was almost unknown among biologists during his day and his results were published in a journal that attracted little attention. This is clearly only part of the story. It is perhaps more correct to say that biologists in 1866 were unable to appreciate the significance of Mendel's work. Their minds were not prepared. There is one interesting bit of information in this connection. Mendel carried on a lengthy correspondence with Nägeli, explaining the results of his experiments. It should be remembered that Nägeli was greatly interested in heredity, he being the proponent of the idioplasm concept. He, of all people, should have seen Mendel's point but he failed to appreciate the significance of the pea experiments.

Still another fact is that Mendel did not believe for long in the universality of his findings. This was the result of an unfortunate choice of material, the hawkweed, for some additional experiments. In the hawkweed the ovules are capable of parthenogenetic development. As a result, many of the crosses he believed he was performing were not crosses at all. Mendel did not realize this and was unable to understand why he did not observe the ratios he had previously found in peas. He probably came to believe that his results held for peas alone.

The three biologists who made the theoretical import of Mendel's paper known to their contemporaries supplied data of their own that could be explained in Mendelian terms, and in the few years after 1900 the results of much more genetic work became available. Most of it could be understood in Mendelian terms, but some could not. Many of the exceptions could not be explained until the physical basis of inheritance was established by cytologists. We should, there- fore, now examine what students of the cell were discovering during the first years of the present century.

SUGGESTED READINGS

Castle, W. C. 1951. 'The beginnings of Mendelism in America.' In *Genetics in the 20th Century,* edited by L. C. Dunn. Macmillan.

Iltis, H. 1932. *Life of Mendel.* Norton.

Mendel, G. (1866). 'Experiments on plant-hybridization.' An English transla- tion of Mendel's paper was published in W. Bateson *Mendel's Principles of Heredity* (Cambridge University Press, 1909, 1913) and in E. W. Sinnott, L. C. Dunn, and Th. Dobzhansky *Principles of Genetics* (McGraw-Hill, 1950). A paper bound edition is available from the Harvard University Press.

Roberts, H. F. 1929. *Plant Hybridization before Mendel.* Princeton Univer- sity Press.

1950. *The Birth of Genetics.* Supplement to Genetics *35.* This consists of Eng- lish translations of the correspondence between Mendel and Nägeli and of the 1900 papers of de Vries, Correns, and Tschermak.

The following genetics textbooks will be found valuable as supplementary reading for the material covered in the remaining chapters on genetics.

King, R. C. 1962. *Genetics.* Oxford University Press.

Sinnott, E. W., L. C. Dunn, and Th. Dobzhansky. 1958. *Principles of Genetics.* McGraw-Hill.

Boveri and Sutton—1902

BOVERI AND THE SEA URCHIN CHROMOSOMES

In Chapter 4, we learned that in 1900 the hypothesis that 'chromosomes are the physical basis of inheritance' seemed to be reasonable. It was far from being established as true and, in fact, there was a real difficulty in knowing how to test such a hypothesis. Until 1902 no one had a clue as to how this might be done. In that year, Boveri carried out an ingenious experiment demonstrating that a complete set of chromosomes was necessary for normal development. Since development is one aspect of inheritance, the relation between chromosomes and inheritance was established.

The significance of Boveri's experiment will be more apparent if we give something of the background from which he worked. At the time of his experiment, most cytologists believed that within any single species one chromosome was about the same as another. Thus, in the sea urchin each cell of the embryo has 36 chromosomes. They were all very small and looked identical in shape; they reacted alike to fixation and staining; and when things look alike there is a natural tendency to believe that they are alike in other respects as well.

Boveri thought otherwise. He believed that chromosomes differed from one another, and that a complete set of 36 was necessary for normal development in the sea urchin. He believed that not just any 36 would suffice: but the very 36 which were present in each cell of the normal embryo were the necessary ones.

Double Fertilization of Sea Urchin Ova. Boveri tested this hypothesis in a clever way. Hertwig and others had observed previously that if one used a highly concentrated sperm suspension, it was possible to get two sperm to enter one egg of the sea urchin. Mitosis becomes most

57

confused in these double fertilizations, but it is possible by this method to vary the number of chromosomes distributed to the cells. In order to understand the complications we should first review normal fertilization.

In the case of an embryo entered by a single sperm, the sperm brings in a *centrosome*, which is the region containing the centriole. (In the sea urchin the centriole is composed of several granules surrounded by an area, the centrosome, which stains differently from the rest of the cytoplasm.) The centrosome divides into two and the spindle forms between the two centrosomes. The paternal and maternal nuclei fuse, and then their chromosomes go on the spindle. Each pronucleus of the sea urchin has 18 chromosomes so the fusion nucleus will have 36. Each of these 36 duplicates itself, thus forming a total of 72 chromosomes. These are divided equally at the first cleavage division, and each daughter cell receives 36.

When *two* sperm enter, not only will there be an additional 18 chromosomes to make a total of 54 but there will be an extra centrosome as well. Boveri observed that one of two things happened:

1. In some embryos the centrosomes of both sperm divided. This resulted in four centrosomes in the single cell. At the time of first cleavage these embryos divided into four instead of into two cells.
2. In the remaining embryos, the centrosome of one sperm divided and the other remained single. This gave three centrosomes for the single cell and at the time of first cleavage this type of embryo divided into three cells.

In both classes of double-fertilization embryos, there would be 54 chromosomes (18 from the maternal nucleus and 18 from each of the two paternal nuclei). These 54 will duplicate themselves during the first mitotic division of the embryo to give a total of 108 chromosomes. In those embryos with four centrosomes, the chromosomes will be divided among four cells. Boveri found this division very unequal. Some cells would get many chromosomes and others only a few. If the apportionment was strictly equal, each cell would receive 27 chromosomes ($108/4 = 27$). This is far from the normal complement of 36 per cell, which Boveri believed necessary for regular development. In fact, there is no way in which each of the four cells could get the normal complement of 36. Abnormal development was to be expected, and this Boveri observed in 1,499 out of 1,500 embryos.

The embryos with three centrosomes that divided into three cells at the first division also showed very abnormal distributions of chromosomes. One would expect, however, that they would have a better chance of getting 36 chromosomes in each cell than would the group that formed four cells. The reason is this: We have seen that there is no way of apportioning 108 chromosomes among four cells so each will receive a normal complement of 36. If the 108 chromosomes are divided equally among three cells, however, the result is 36. The experimental results validated this reasoning. In the group that divided into three cells, 58 in a total of 719 developed normally. We have already seen that only one embryo in 1,500 developed normally among the embryos that divided into four cells at first cleavage.

According to Boveri, these data correspond fairly well with the chance expectation that normal larvae will come from embryos that begin development with a normal set of chromosomes in each of the cells formed at the first division. He interpreted the data to mean that for normal development every cell of the embryo must have the regular set of 36 chromosomes. He believed that every single chromosome in the set must be endowed with a different quality, and that all are necessary for normal development.

These experiments emphasized the importance of chromosomes for normal development, which is one aspect of inheritance. It was a direct approach to the study of the role of chromosomes in inheritance.

SUTTON AND GRASSHOPPER CHROMOSOMES

In the same year that Boveri published the results of his work, a second and much more fruitful approach was made by Sutton. At the time, he was a graduate student working at Columbia University with the cytologist Wilson. He published two papers on the chromosomal basis of inheritance, the first in 1902 and the second in 1903.

Individuality of the Chromosomes. Sutton's 1902 paper was a study of the chromosomes in the testis of a grasshopper of the genus Brachystola. The chromosomes of this form

exhibit a chromosome group, the members of which show distinct differences in size. Accordingly one feature of this study has been a critical examination of large numbers of dividing cells (mainly from the testis) in order to determine whether, as has usually been taken for granted, these differences are merely a matter of chance, or whether in accordance with the view recently expressed by Montgomery, . . . characteristic size relations are a constant attribute of the chromosomes individually considered. With the aid of camera drawings of the chromosome group in the various cell-generations, I will give below a brief account of the evidence which has led me to adopt the latter conclusion.

The cells in the testis undergo a series of mitotic divisions before they begin meiosis. These cells are known as *spermatogonia* and they have the diploid number of chromosomes. The youngest spermatogonia that Sutton could find possessed 23 chromosomes. One of these, the 'accessory' chromosome, had a peculiar behavior and it will be considered separately. The other 22 were of various sizes. When these were measured carefully it was found that there were not 22 different sizes, but only 11. In other words there were two chromosomes of each size class. In addition to the minor size variation, the chromosomes could be divided into two groups that differed strikingly in size. Three of the pairs were very small and the other eight were large.

Sutton found that these early spermatogonia went through eight mitotic divisions. At each metaphase the same 11 pairs of chromosomes were observed. Of these, eight pairs were large and three small. He concluded that constant size was an attribute of the individual chromosomes.

After these eight mitotic divisions, the cells undergo the usual two meiotic divisions. The chromosomes synapse in pairs, each member of the pair being of the same size. As a result, 11 tetrads form; eight are large and three are small. Then the two meiotic divisions occur and each sperm receives one chromosome of each of the 11 sizes.

Sutton found that the diploid number in the female was 22. Further, these had the same size relations as were present in the male, consisting of eight large and three small pairs. (Sutton made an error in this count. Later workers found 24 chromosomes.) He postulated that every ovum after meiosis would contain one chromosome of each of the 11 sizes.

Fertilization of an ovum containing 11 chromosomes with a sperm containing 11 would restore the diploid number of 22. Some of the sperm will have an accessory chromosome in addition to the regular 11. Fertilization with a sperm of this type will result in a zygote with 22 chromosomes plus an accessory. (In 1901 McClung suggested that the accessory chromosome is in some way concerned with sex determination. More will be said about this in Chapter 7.) Sutton continues:

Taken as a whole, the evidence presented by the cells of Brachystola is such as to lend great weight to the conclusion that a chromosome may exist only by virtue of direct descent by longitudinal division from a preëxisting chromosome and that the members of the daughter group bear to one another the same respective relations as did those of the mother group—in other words, that the chromosome in Brachystola is a distinct morphological individual.

This conclusion inevitably raises the question whether there is also a physiological individuality, i.e., whether the chromosomes represent respectively different series or groups of qualities or whether they are merely different-sized aggregations of the same material and, therefore, qualitatively alike.

On this question my observations do not furnish direct evidence. But it is *a priori* improbable that the constant morphological differences we have seen should exist except by virtue of more fundamental differences of which they are an expression; and, further, by the unequal distribution of the accessory chromosome we are enabled to compare the developmental possibilities of cells containing it with those of cells which do not. Granting the normal constitution of the female cells examined and the similarity of the reduction process in the two sexes, such a comparison must show that this particular chromosome does possess a power not inherent in any of the others—the power of impressing on the containing cell the stamp of maleness, in accordance with McClung's hypothesis.

The evidence advanced in the case of the ordinary chromosomes is obviously more in the nature of suggestion than of proof, but it is offered in this connection as a morphological complement to the beautiful experimental researches of Boveri already referred to. In this paper Boveri shows how he has artificially accomplished for the various chromosomes of the sea-urchin, the same result that nature is constantly giving us in the case of the accessory chromosome of the Orthoptera. He has been able to produce and to study the development of blastomeres lacking certain of the chromosomes of the normal series.

If, as the facts in Brachystola so strongly suggest, the chromosomes are persistent individuals in the sense that each bears a genetic relation to one only of the previous generation, the probability must be accepted that each represents the same qualities as its parent element. A given relative size may, therefore, be taken as characteristic of the physical basis of a certain definite set of qualities. But each element of the chromosome series of the spermatozoon has a morphological counterpart in that of the mature egg and from this it follows that the two cover the same field in development. When the two copulate, therefore, in synapsis the entire chromatin basis of a certain set of qualities inherited from the two parents is localized for the first and only time in a single continuous chromatin mass; and when in the second spermatocyte division, the two parts are again separated, one goes entire to each pole contributing to the daughter cells the corresponding group of qualities from the paternal or the maternal stock as the case may be.

There is, therefore, in Brachystola no qualitative division of chromosomes but only a separation of the two members of a pair which, while coexisting in a single nucleus, may be regarded as jointly controlling certain restricted portions of the development of the individual. By the light of this conception we are enabled to see an explanation of that hitherto problematical process, synapsis, in the provision which it makes that the two chromosomes representing the same specific characters shall in no case enter the nucleus of a single spermatid or mature egg.

I may finally call attention to the probability that the association of paternal and maternal chromosomes in pairs and their subsequent separation during the reducing division as indicated above may constitute the physical basis of the Mendelian law of heredity. To this subject I hope soon to return in another place.

The Chromosomes in Heredity. The far-reaching suggestion made in the last paragraph was developed in Sutton's 1903 paper entitled 'The Chromosomes in Heredity.' In this he pointed out that the segregation and recombination of genes as studied by the geneticists showed a striking parallel to the behavior of chromosomes as revealed by the cytologists. The pertinent cytological data, according to Sutton, were as follows:

1. The diploid chromosome group consists of two morphologically similar chromosome sets. Every chromosome type is represented twice. Expressed another way, chromosomes exist in homologous pairs. Strong grounds exist for the belief that one set was derived from the father and one set from the mother at the time of fertilization.

2. Synapsis consists of the pairing of homologous chromosomes.

3. As a result of meiosis every gamete receives only one chromosome of each homologous pair.

4. The chromosomes retain their morphological individuality throughout the various cell divisions.

5. The distribution in meiosis of the members of each homologous pair of chromosomes is independent of that of each other pair. As a result, each gamete receives a random assortment of chromosomes so far as the members of each pair are concerned.

Sutton then made the point that Mendel's results could be explained on the assumption that genes were parts of the chromosomes. The following example will show how this is possible:

Let us assume that the *round* and *wrinkled* genes in peas are carried on a certain pair of chromosomes (Fig. B6-1). If a chromosome has the *round* gene, we shall call it **R** and if it has the *wrinkled* gene, we shall call it **r**. Let us further assume that the *yellow* and *green* genes are carried by a different pair of chromosomes. If the chromosome has the *yellow* gene, we shall designate it **Y** and if it carries the *green* gene, we shall call it **y**. A pure-breeding *round-yellow* plant would be symbolized as **RRYY** indicating that it had a pair of chromosomes carrying the *round* gene and another pair with the *yellow* gene. Similarly, a *wrinkled-green* plant would be **rryy**.

When the reduction divisions occur the *round-yellow* plant would produce haploid gametes with one **R** and one **Y** chromosome—or **RY**. This alone could result, since a gamete receives only one chromosome of every kind. A **RR** or a **YY** gamete would be impossible in normal

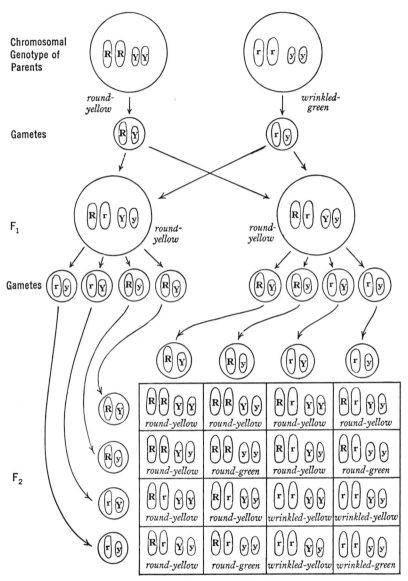

6-1 Diagram of chromosome distributions in Mendel's cross of a *round-yellow* X *wrinkled-green* pea on the basis of Sutton's hypothesis.

meiosis. The *wrinkled-green* plant would produce gametes solely of the ry type. A union of gametes of the two plants would result in one type of offspring, namely, **RrYy**. This F_1 individual would be diploid and would have received one member of each chromosome type from the male gamete and one from the female gamete.

The gametes of the F_1 plant would be of four possible types. The **R** and the **r** would go to different cells during meiosis. The **Y** and **y** chromosomes would likewise be separated and *their separation would not affect the separation of* **R** *and* **r**. Thus, all possible combinations, namely, **RY**, **Ry**, **rY**, and **ry** would be produced, and in approximately equal numbers. If you will re-examine Mendel's description of this cross (p. 194), you will note the exact parallel between the scheme for Mendel's breeding experiments and the chromosome movements just described. The F_2 chromosomes would be of the type shown in the genetic checkerboard. The sole difference to be noted is that Mendel characterized his pure-breeding peas of the parental generation as **RY** and **ry**, while we have used a diploid chromosome designation **RRYY** and **rryy**. Mendel could have used **RRYY** and **rryy** just as well.

Sutton's Hypothesis. We may conclude, therefore, that genes of a type postulated by Mendel could be

1. parts of the chromosomes, or
2. parts of some other cell structures that behave in the same way as chromosomes in mitosis, meiosis, and fertilization.

When a scientist is confronted with two hypotheses, one involving known factors and the other invoking unknown factors, the first is usually chosen. In the case under consideration, such a choice would have a great practical advantage: It would be easier to make observations and design experiments to test the role of chromosomes in Mendelian heredity than it would be to investigate the role of some unknown cell structures.

Sutton's general hypothesis was not new. As we have already seen, some cytologists believed, at least as early as 1884, that the chromosomes were involved in inheritance. Sutton pointed out additional reasons for so thinking and, even more important, made a definite link between genetic data and cytological data. If we continue to use the genes-are-parts-of-chromosomes hypothesis, it will be necessary to find a parallel between all types of genetic behavior and chromosome behavior. Any variations in chromosomal phenomena from the usual condition must be reflected in the genetic results. Similarly, if genetic ratios are obtained that cannot be explained in Mendelian terms, one must find a chromosomal basis for the deviation. Sutton indicated one type of genetic behavior that could be expected if his hypothesis was correct.

We have seen reason, in the foregoing considerations, to believe that there is a definite relation between chromosomes and allelomorphs or unit characters, but we have not before inquired whether an entire chromosome or only a part of one is to be regarded as the basis of a single allelomorph. The answer must unquestionably be in favor of the latter possibility, for otherwise the number of distinct characters possessed by an individual could not exceed the number of chromosomes in the germ-products; which is undoubtedly contrary to fact. We must, therefore, assume that some chromosomes at least are related to a number of different allelomorphs. If then, the chromosomes permanently retain their individuality, it follows that all the allelomorphs represented by any one chromosome must be inherited together.

If Sutton's reasoning is correct, the Mendelian principle of independent assortment could apply only to cases where the two pairs of contrasting factors were carried on separate chromosomes.

The importance of Sutton's theoretical considerations can scarcely be overemphasized. Two completely different disciplines were found to have an area in common: cytology and genetics became mutually supporting and ˙stimulating fields. Theories of inheritance could be 'double-checked.'

SUGGESTED READINGS

Sutton, W. S. 1902. 'On the morphology of the chromosome group in *Brachystola magna.' Biological Bulletin 4*:24–39.
Sutton, W. S. 1903. 'The chromosomes in heredity.' *Biological Bulletin 4*:231–51.

7

Sex Chromosomes

The first impression cytologists gained from a study of chromosomes was a feeling that the behavior of these cell structures was similar in all animals. Whether the form studied was a worm, snail, salamander, or mammal, one observed the same duplication of each chromosome to form two chromatids during mitosis. This was followed by the distribution of one of the two chromatids to each daughter cell. Each chromosome appeared double at the same time, and the movements during metaphase, anaphase, and telophase were synchronous. A similarity of behavior was observed in the events connected with the reduction of chromosome number in meiosis of both male and female gametes. It should be emphasized that this similarity of behavior applied not only to the behavior of the group of chromosomes but to the individual chromosomes in the group as well.

Atypical Chromosomes. It was not long before this concept of uniform behavior was found to have its exceptions. In the last decade of the nineteenth century and the first years of the twentieth, a few cases were reported of one or two of the chromosomes of a set behaving in a manner quite unlike the rest. The unusual behavior might be a difference in reaction of the chromosome to stains, meiotic movements that were not synchronous, or the presence of 'extra' or 'accessory chromosomes.' The term *accessory chromosome* referred to those cases where there was one chromosome without a mate, in contrast to the usual situation of all chromosomes being in morphologically similar pairs.

The analysis of accessory chromosomes led to important results concerning the role of chromosomes in heredity. As is frequently the case in science, these observations were made and recorded before their true significance was understood.

Henking's Description of the **X** *Chromosome.* In 1891 Henking published his observations on chromosome behavior during sperm formation in a bug, Pyrrhocoris (Fig. 7–1). This species has 23 chromosomes. Twenty-two of them form 11 pairs, the two members of a pair having the same appearance. The extra chromosome was called **X**. It did not have a mate. During the first meiotic division the 22 chromosomes synapsed to form 11 pairs. Later these formed tetrads.

The behavior of the **X** chromosome was different. Having no mate, it could not synapse and form a tetrad. It did duplicate itself, however, to form a structure like a dyad. At the beginning of meiosis, the cells therefore contained 11 tetrads, plus the **X** in the form of a dyad. At the first meiotic division the 11 tetrads were separated but the **X** went entire to one of the daughter cells. At the end of the first division, one of the daughter cells contained 11 dyads of the usual sort, plus the **X** dyad. The other daughter cell contained only the 11 dyads. At the second meiotic division of the cell with the **X**, the **X** dyad and the 11 regular dyads were divided so each of the resulting cells contained an **X** chromosome plus 11 of the regular chromosomes. In the cell without the **X** dyad, the other dyads were divided so each daughter cell contained 11 regular chromosomes. Therefore, the four cells resulting from the two meiotic divisions consisted of two with 11 chromosomes plus an **X**, and two with the 11 chromosomes alone. So far as the **X** was concerned, we could say that half of the sperm contained an **X** and the other half did not.

During the next decade, many workers discovered these chromosomes with atypical behavior. They were given a variety of names such as 'X chromosomes' and 'accessory chromosomes.' In every case the accessory was unique in some feature such as stainability, time of movement to the poles of the spindle, enclosure in a separate vesicle instead of the nucleus, lack of a mate during synapsis, or distribution to only half of the sperm.

Sex and the Accessory Chromosomes. In 1901 McClung suggested that the accessory chromosome was in some way connected with sex determination:

Being convinced from the behavior in the spermatogonia and the first spermatocytes of the primary importance of the accessory chromosome, and attracted by the unusual method of its participation in the spermatocyte mitoses, I sought an explanation that would be commensurate with the importance of these facts. Upon the assumption that there is a qualitative difference between the various chromosomes of the nucleus, it would necessarily follow that there are formed two kinds of spermatozoa which, by fertilization of the egg, would produce individuals qualitatively different. Since the number

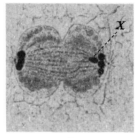

a

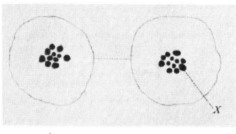

b   c

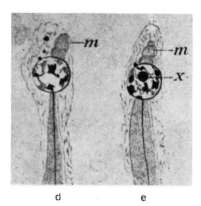

d   e

7-1 Meiosis in Pyrrhocoris. *a* shows a cell in the telophase of the second meiotic division. Since this is a lateral view not all of the chromosomes are shown. The peculiar body, indicated as 'X' by Henking, goes to one pole of the spindle and will, therefore, be in only half of the cells formed by this division. Consequently two types of cells result from meiosis. One type, *b*, has 11 chromosomes and the other type, *c*, has 11 chromosomes plus the X. These cells develop directly into sperm. Thus, half of the sperm will have an X (*e*) and the others lack an X (*d*) (H. Henking, 'Untersuchungen über die ersten Entwicklungsvorgänge in den Eiern der Insekten,' *Zeit. für wiss. Zool.* 51:685–736. 1891).

of each of these varieties of spermatozoa is the same, it would happen that there would be an approximately equal number of these two kinds of offspring. We know that the only quality which separates the members of a species into these two groups is that of sex. I therefore came to the conclusion that the accessory chromosome is the element which determines that the germ cells of the embryo shall continue their development past the slightly modified egg cell into the highly specialized spermatozoon.

It would not be desirable in a preliminary paper of this character to extend it by a detail of the discussion by which the problem was considered. Suffice it to say that by this assumption it is possible to reconcile the results of many empirical theories which have proved measurably true upon the general ground that the egg is placed in a delicate adjustment with its environment, and in response to this, is able to attract that form of spermatozoon which will produce an individual of the sex most desirable to the welfare of the species. The power of selection which pertains to the female organism is thus logically carried to the female element.

Numerous objections to this theory received consideration, but the proof in support of it seemed to overbalance them largely, and I was finally induced to commit myself to its support. I trust that the element here discussed will attract the attention which I am convinced it deserves and can only hope that my investigations will aid in bringing it to the notice of a larger circle of investigators than that now acquainted with it.

McClung's hypothesis was not accepted or even widely believed at first. In part this was due to the type of reasoning displayed in the second paragraph of the quotation. It was difficult to imagine how an unfertilized ovum could select the type of sperm and so produce the 'sex most desirable to the welfare of the species.' Were this possible, it would be a most interesting extension of feminine intuition! The question was all the more confusing when somewhat later it was found that the female had not one less chromosome than the male, but one more!

Clarification of Chromosomal Sex Determination. In 1905 the situation was clarified by Wilson and one of his students, Stevens. They studied meiosis in a number of insects and found that **X** chromosomes were the rule, not the exception. Because of the importance of **X** chromosomes in sex determination they were called *sex chromosomes*. All other chromosomes were called *autosomes*. Thus, Henking's bug, Pyrrhocoris, would have one **X** sex chromosome and 22, or 11 pairs, of autosomes.

Stevens and Wilson found two types of sex chromosome behavior, the **X0-XX** type and the **XY-XX** type (Fig. 7–2).

The **X0-XX** *Type.* In the species having this type of sex chromosome behavior, the male has a single **X** chromosome and the female has two **X** chromosomes. The male can be symbolized as **X0**, where **0** signifies

XO-XX TYPE OF SEX DETERMINATION

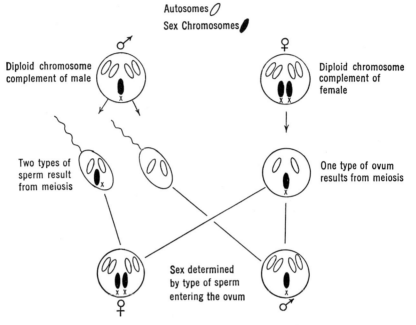

XY-XX TYPE OF SEX DETERMINATION

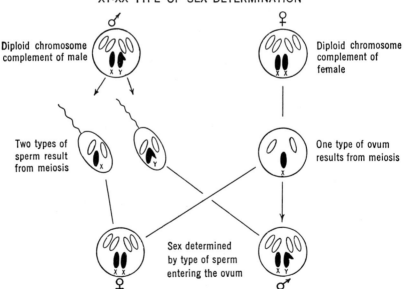

7–2

the absence of a homologue of the **X**. The female is **XX**. Both sexes will have the same autosomes. A Pyrrhocoris male would have 11 pairs of autosomes and one **X** chromosome. In meiosis two types of sperm will be produced: One type will contain 11 autosomes plus the **X** and the other type will contain 11 autosomes and no **X**. The ova will all be of one type, containing 11 autosomes and one **X**. The union of a sperm of the first type with an ovum will result in a zygote with 22 autosomes and two **X** chromosomes. This individual will be a female. Fertilization of an ovum with the second type of sperm will result in a zygote with 22 autosomes and a single **X** (that is, **X0**). The result will be a male.

The **XY-XX** *Type.* In this type, both the male and female possess a pair of sex chromosomes. Once again, the female has a pair that are identical and she is symbolized as **XX**. The male has one chromosome that is identical with the **X** of the female plus another, the **Y**, that is morphologically different. The **Y** might be longer, shorter, or of a different shape. It is similar to the **X** in some way, since synapsis occurs between **X** and **Y**.

The results of meiosis would be these: Every ovum would contain autosomes (the number depending on the species) and one **X**. The sperm would be of two types. One would contain autosomes and one **X**, and the other would contain autosomes and one **Y**. The fertilization of an ovum with an **X**-bearing sperm would give a zygote with the diploid set of autosomes and **XX**. This would be a female. The fertilization of an ovum with a **Y**-bearing sperm would give a zygote with the diploid set of autosomes and **XY**. This would be a male. (Incidentally, man has the **XX-XY** type of sex determination and so does *Drosophila melanogaster,* a small fly much used in genetic work.)

These two types of sex chromosome distribution described by Stevens and Wilson are those most frequently encountered in the animal kingdom. With each type, the male produces two classes of sperm, and sex is determined by the class of sperm entering the ovum. Additional types were discovered later. In birds, for example, it is the female that produces two classes of gametes and the male only one.

If these observations on sex chromosomes are correct, we may draw some important conclusions:

1. Sex is determined at the time of fertilization.
2. If sex determination is due only to sex chromosomes, we can regard the sex of an individual as irreversible, unless we can alter the chromosomes.

3. The two sexes should be produced in approximately equal numbers. (Why?)

4. The relation between sex and chromosomes is additional evidence supporting Sutton's hypothesis that chromosomes are the basis of inheritance.

With the data already presented, let us speculate a bit, using sex determination in species of the **XY**-male–**XX**-female type as the basis of our speculations: The constant association of a distinctive chromosome type with each sex suggests a causal relation. What is the relation? Is the male a male because he has a **Y**, or because he has only one **X**, or because he has both an **X** and a **Y**? Do other possibilities suggest themselves? Try to imagine how you could test these various possibilities.

SUGGESTED READINGS

Wilson, E. B. 1928. *The Cell in Development and Heredity.* Macmillan. Chapter 10, 'Chromosomes and Sex.'

Variations in Mendelian Ratios

One of the most striking things about Mendel's results was the uniform behavior in inheritance of the seven pairs of contrasting characters used in his crosses. One member of each pair was always dominant and the other always recessive. In a cross of a pure-breeding dominant and a pure-breeding recessive, the F_1 population consisted only of plants with the dominant characteristics. The F_2 of such a cross consisted of both of the original types, in the ratio of three dominants to one recessive. If one of the F_1 plants was crossed with a pure recessive, the ratio of dominants to recessives in the offspring was $1:1$. If two pairs of contrasting characters were involved, the F_2 ratio was $9:3:3:1$. If the F_1 of the cross just mentioned was crossed to a double recessive, the ratio was $1:1:1:1$.

Confirmations and Exceptions to Mendel's Scheme. In the first few years after 1900, when Mendel's findings became widely known to biologists, the results of many crosses were reported. The majority of these gave ratios that were nearly identical with Mendel's pea crosses. A few gave slightly different results that could be understood with minor adjustments of the Mendelian scheme. Still others defied explanation at that time. This last group of cases was put aside in the hope that eventually a modified Mendelian theory could explain the results. Subsequent events showed this to be a profitable procedure.

In these early years of the twentieth century, the British geneticist, Bateson, was the most vigorous disciple of Mendelism. Even before Mendel's work was known to the scientific world at large, Bateson had undertaken an extensive program of breeding. As a result of this preparation, he was in a proper frame of mind to grasp the significance of Mendel's work. In England, Bateson waged a scientific battle to convince his fellow biologists that Mendel's approach to the problem of inheritance was a useful one. This 'battle' was not always waged with the fairness, objectivity, lack of bias, and honest criticism that should be the basis for scientific discussions. In fact, the participants behaved like ordinary human beings. Much of the genetic work of this period was summarized in reports Bateson made to the Royal Society (1902, 1905, 1906, 1908, 1909) and in his book, *Mendel's Principles of Heredity* (1909).

Some Useful Genetic Terms. Bateson and others introduced some terms that are useful in describing genetic events. From Mendel's experiments he 'reached the conception of unit-characters existing in antagonistic pairs. Such characters we propose to call *allelomorphs* [now usually called *alleles*], and the zygote formed by the union of a pair of opposite allelomorphic gametes, we shall call a *heterozygote*. Similarly, the zygote formed by the union of gametes having similar allelomorphs, may be spoken of as a *homozygote*.'

To use these terms in examples, we might observe that in peas *round* and *wrinkled* genes are alleles; so are *yellow* and *green* genes. A pure-breeding *round* plant, which we have designated **RR**, is homozygous. A *wrinkled* plant, **rr**, is also homozygous. The F_1 of a cross of these is **Rr** and heterozygous. Recalling that *round* is dominant, we know that a *round* plant can be either homozygous (**RR**) or heterozygous (**Rr**), and that one could not tell by external appearance which was which. It is worth repeating that there is a difference between appearance and genetic make-up. We speak of the appearance of the organism as its *phenotype* (*yellow* or *green*) and the genetic make-up as its *genotype* (**RR**, **Rr**, or **rr**). The individual factors of inheritance, **R** or **r**, carried on the chromosomes are the *genes*.

Returning now to Bateson, these are some of the types of genetic data he gave:

EXCEPTIONS TO THE MENDELIAN RATIOS

1. Blended Character Expression in Heterozygotes. In Mendel's results the heterozygote was always identical in appearance with the homozygous dominant. Even today we recognize this as the most frequent condition, but there are some cases where the heterozygote is intermediate. In such a situation neither allele is dominant or recessive. One example is a common cultivated flower, the four o'clock. When a *red*-flowered four o'clock is crossed with a *white*-flowered one, all of the F_1 plants have *pink* flowers. In the F_2, *red-*, *white-*, and *pink-*flowered plants appear. The genetic basis of this cross is diagrammed below:

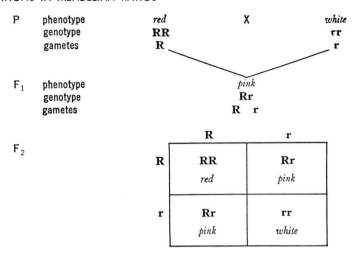

P	phenotype	*red*	X	*white*
	genotype	**RR**		**rr**
	gametes	**R**		**r**

F₁	phenotype	*pink*
	genotype	**Rr**
	gametes	**R r**

F₂

	R	r
R	**RR** *red*	**Rr** *pink*
r	**Rr** *pink*	**rr** *white*

F₂ ratio: 1 *red*, 2 *pink*, 1 *white*.

2. Two Pairs of Alleles Affecting the Same Character. In all the crosses considered so far, a character has been affected by only one pair of alleles. This may seem to imply that every character of an organism is determined by a single pair of alleles. This is not the case, as shown by some experiments with chickens which were reported by Bateson.

Poultry breeders recognized a number of comb types, usually involving differences in comb shape. Some of these types were called *single, rose, pea,* and *walnut.* A cross between *rose* and *single* gave *rose* in the F₁ and 3 *rose* to 1 *single* in the F₂. *Rose* and *single* behaved as ordinary alleles with *rose* the dominant.

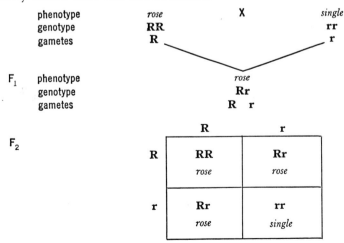

	phenotype	*rose*	X	*single*
	genotype	**RR**		**rr**
	gametes	**R**		**r**

F₁	phenotype	*rose*
	genotype	**Rr**
	gametes	**R r**

F₂

	R	r
R	**RR** *rose*	**Rr** *rose*
r	**Rr** *rose*	**rr** *single*

Similarly, a cross between *pea* and *single* gave *pea* in the F_1 and 3 *pea* to 1 *single* in the F_2. *Pea* and *single* behaved as ordinary alleles with *pea* the dominant.

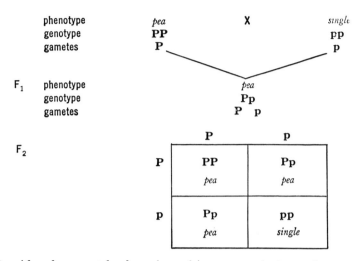

phenotype	*pea*	X	*single*
genotype	**PP**		**pp**
gametes	**P**		**p**

F_1

phenotype	*pea*
genotype	**Pp**
gametes	**P p**

F_2

	P	**p**
P	**PP** *pea*	**Pp** *pea*
p	**Pp** *pea*	**pp** *single*

Considered separately there is nothing unusual about these crosses. Considered together there is one puzzle, largely semantic: why is the genotype of *single* written as **rr** in the first cross and as **pp** in the second? The answer to this question hinges on the fact that the *single* comb condition results from the interaction of two different genes. The two previous crosses and the next one to be described show that both the **r** and **p** genes are involved in the genetic determination of comb shape (subsequent experiments showed that additional genes were concerned). A comb of the *single* type forms in a chicken when both the *pea* and *rose* genes are recessive. The genotype of a *single* comb chicken, therefore, is **pprr**. A *pea* comb chicken would have the genotype **PPrr** or **Pprr** and a *rose* comb chicken would have the genotype **ppRR** or **ppRr**. We might have written the cross of *rose* comb × *single* comb as **ppRR** × **pprr**. At that time, however, we were unaware of the existence of the **p** genes. Furthermore, there was no need to do so since both parents are of the same genotype with respect to the **p** gene: by convention geneticists use the symbols only for those cases where the genes differ in the two parents. If the cross involves differences in both the **p** and **r** comb shape genes, then the necessary genetic symbols must be employed. Such a cross will now be described.

Let us cross the two dominants, *rose* and *pea*. The F_1 will be found to have a new type of comb, *walnut*, which might be looked upon as a blending of the two dominant shapes. The F_2 will give a ratio of 9 *walnut, 3 rose, 3 pea,* and 1 *single.*

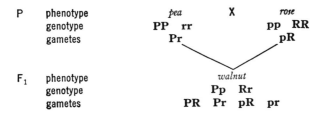

	PR	Pr	pR	pr
PR	PP RR *walnut*	PP Rr *walnut*	Pp RR *walnut*	Pp Rr *walnut*
Pr	PP Rr *walnut*	PP rr *pea*	Pp Rr *walnut*	Pp rr *pea*
pR	Pp RR *walnut*	Pp Rr *walnut*	pp RR *rose*	pp Rr *rose*
pr	Pp Rr *walnut*	Pp rr *pea*	pp Rr *rose*	pp rr *single*

F_2 ratio - 9 *walnut;* 3 *rose;* 3 *pea;* 1 *single.*

The *single* comb is the result of both (recessive) **p** and **r** genes being homozygous. *Rose* is obtained if the animal is homozygous or heterozygous for **R** and homozygous for **p**. *Pea* is obtained if the animal is homozygous or heterozygous for **P** and homozygous for **r**. *Walnut* results when there is at least one **P** gene together with at least one **R** gene.

3. Yellow Mice. Cuénot reported some crosses in mice that neither he nor Bateson could explain in the usual Mendelian manner. He worked with *yellow* and *agouti* hair color genes. These two types of crosses will demonstrate the problem.

1. *yellow* $\times$ *yellow* gives 2 *yellow* to 1 *agouti.*

2. *yellow* $\times$ *agouti* gives 1 *yellow* to 1 *agouti.*

Cuénot found it impossible to obtain a strain of *yellow* mice that would breed true. The *yellow* animals always behaved as though they were heterozygous. His *agouti* strains bred true. During the course of his experiments he observed that the litter size in crosses of *yellow* × *yellow* was smaller than in the *yellow* × *agouti* cross.

Can you devise a hypothesis to explain the results? From the data given it is possible to arrive at the explanation that was later found to be correct.

4. Coupling of Genes. Bateson listed another type of result that could not be explained, namely, cases involving two pairs of genes that did not show independent assortment. Let us assume that there are two pairs of genes, **A** and **a** and **B** and **b** and that the phenotypic expression of **A** is *A*, that of **a** is *a*, and so on. In the cross **AABB** × **aabb** we should expect an F₂ phenotypic ratio of 9 *AB*, 3 *Ab*, 3 *aB*, and 1 *ab* if Mendel's rules applied. In some crosses, however, Bateson found that the *A* and *B* characters appeared to be completely *coupled*, that is, they were inherited together. The same was true for *a* and *b*. In fact, the F₂ ratios might be close to 3 *AB* to 1 *ab* with few or no individuals of the *Ab* or *aB* phenotypes. Results of this sort could not be accounted for by the Mendelian scheme but they had been predicted by Sutton (Chapter 6). The solution to these exceptional cases will be given in Chapter 10.

SUGGESTED READINGS

Bateson, W. 1909. *Mendel's Principles of Heredity.* Cambridge University Press.
Punnett, R. C. 1950. 'The Early Days of Genetics.' *Heredity 4*:1–10.

Morgan's White-eyed Drosophila

In this chapter we shall begin a survey of some of the most important advances in genetics. During the first ten years following the rediscovery of Mendel's experiments, the progress of genetics was slow though steady. It was found that Mendel's scheme worked for many organisms and not for peas alone. To be sure some crosses gave ratios that were different from those expected in the Mendelian scheme. These proved difficult to analyze. During this same decade Sutton had suggested that the chromosomes might provide the physical basis for inheritance but those biologists concerned with breeding experiments were unable to appreciate the force of his arguments and data.

In 1910 the American geneticist, Thomas Hunt Morgan, together with his students A. H. Sturtevant, C. B. Bridges, and H. J. Muller, began a remarkable series of experiments. In a period of one decade their efforts changed genetics into the most systematized, so far as concepts and data were concerned, of all branches of biology.

It sometimes appears that much of the progress in science is due to fortunate accidents. One of these accidents was the choice by Morgan of the small fly, *Drosophila melanogaster,* for genetic work. If Drosophila had never been used, the progress of genetics would have been very much slower. Even to this day no other species so useful has been discovered.

The Origin of Hereditary Variation. Morgan began his work with Drosophila to answer a question about the origin of hereditary variation in organisms. Up to this point we have discussed alleles without reference to their possible origin. One of the pairs of alleles that Mendel used was the **R** and **r** genes, which determined whether the pea seeds would be *round* or *wrinkled*. Mendel obtained the seeds that he used from seed dealers, who in turn probably obtained them originally from farmers. What was the origin of the round and wrinkled varieties? Were all peas originally round and then did one suddenly become wrinkled? Or perhaps it was the reverse.

The Mutation Theory of de Vries. In 1901–3 the Dutch botanist, Hugo de Vries, from his studies of the evening primrose (Oenothera) advanced the hypothesis that abrupt changes can occur in the hereditary material of an organism. He believed that these changes were fairly frequent and once they had occurred they tended to be inherited. De Vries would have maintained that the gene causing *round* seeds in peas could change to a gene that caused *wrinkled* seeds. He spoke of this process of change as *mutation* and the new variety as a *mutant.*

The White-eyed Mutant. Morgan began raising Drosophila in the hope of observing the origin of mutants. The first mutant that he found was a male with *white* eyes. It suddenly appeared in a culture bottle of *red*-eyed flies. *Red* is the 'normal' or 'wild type' color of the eyes in *Drosophila melanogaster.* He began experiments to determine the mode of inheritance of the *white*-eyed condition.

In the first cross Morgan mated the *white*-eyed mutant male with a *red*-eyed female (the symbols ♂ and ♀ are used for male and female respectively). These were the results including the actual numbers of individuals obtained in the F_2.

P	*red* ♀	X	*white* ♂
F_1	*red* ♀	*red* ♂	

F_2	*red* ♀	*white* ♀	*red* ♂	*white* ♂
	2,459	0	1,011	782

In addition, the original *white* male was crossed to one of his F_1 daughters.

P	*red* ♀ (F₁ above)	X	*white* ♂

F_1	*red* ♀	*white* ♀	*red* ♂	*white* ♂
	129	88	132	86

Explaining the Cross: First Hypothesis. These results could not be explained by the usual Mendelian scheme. The peculiar relation of eye color to sex, with the absence of *white* females in the F_2 of the first cross, suggested that sex chromosomes might be involved. Morgan proposed the following hypothesis to explain the data:

Let us call the gene that results in *white* eyes, **w**, and the gene that results in *red* eyes, **W**. The *white*-eyed male will produce sperm, all of which will carry **w**. Half of these sperm will have, in addition, an **X** chromosome; the other half will not. The genotype of the original *white* male could be written **wwX**. Two types of sperm will be produced, namely, **wX** and **w**. The *red*-eyed female would be **WWXX**. These symbols represent the two *red* genes and the two **X** chromosomes. All of the eggs would be **WX**.

Morgan's first cross could be represented by the following scheme.

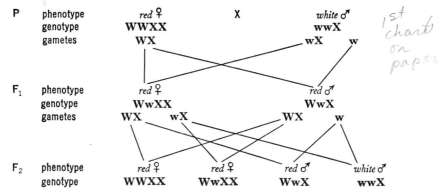

This scheme fits the experimental results, but Morgan pointed out that it is necessary to make one assumption about gamete formation in the F_1 *red* males. In these males, which are heterozygous for *white* eyes, it is necessary to assume that the **W** gene always goes to the same pole of the spindle with the **X** and that the **w** gene never does during the chromosome movements of meiosis. Consequently, there are no **wX** gametes formed by **WwX** males. He adds, 'This all-important point can not be fully discussed in this communication.'

If Morgan's theory was correct, it should have been possible to make deductions about the behavior of the various genotypes and to test these deductions experimentally. Morgan made four such deductions and tested them by making the appropriate crosses.

1. If the genotype of the *white* male is **wwX** and of the *white* female **wwXX**, the following would be expected in a cross of these two types:

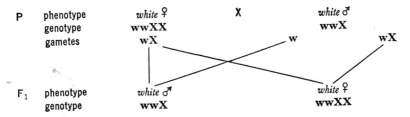

This cross was made and the results were entirely according to expectation, that is, only *white*-eyed flies were obtained.

2. The hypothesis requires that two genotypes be present in the F_2 females, namely, **WWXX** and **WwXX**. The two types could be differentiated by crossing to *white* males. These results would be expected:

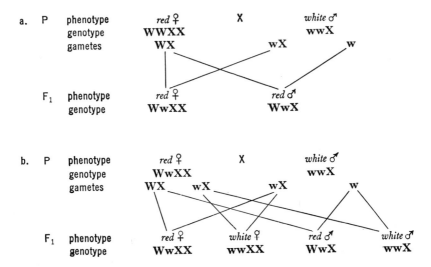

a. P phenotype   *red* ♀   X   *white* ♂
       genotype   **WWXX**   **wwX**
       gametes   **WX**   **wX**   **w**

    F_1 phenotype   *red* ♀   *red* ♂
           genotype   **WwXX**   **WwX**

b. P phenotype   *red* ♀   X   *white* ♂
       genotype   **WwXX**   **wwX**
       gametes   **WX** **wX**   **wX**   **w**

    F_1 phenotype   *red* ♀   *white* ♀   *red* ♂   *white* ♂
           genotype   **WwXX**   **wwXX**   **WwX**   **wwX**

Tests of the F_2 *red* females showed in fact that these two classes exist.

3. The genotype of the F_1 female in the original cross was thought to be **WwXX**. If this was correct, a cross of the F_1 female and a *white* male would give the same results as cross 2b (above). This cross was made and the prediction verified.

4. The hypothesis requires the F_1 male in the original cross to be **WwX**. If such a male is crossed to a *white* female, the following results would be expected:

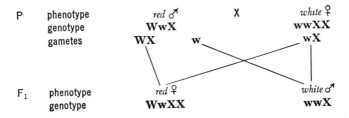

P phenotype   *red* ♂   X   *white* ♀
   genotype   **WwX**   **wwXX**
   gametes   **WX**   **w**   **wX**

F_1 phenotype   *red* ♀   *white* ♂
       genotype   **WwXX**   **wwX**

Once again, the actual experiment yielded the expected results. Note, however, the assumption that **W** and **X** were always together in the same sperm and that no **wX** sperm were formed by **WwX** males.

Nearly all of Morgan's hypothesis was based on facts or on ideas that seemed quite probable. The role of chromosomes in sex determination, the concept of alleles showing dominance and recessiveness, and the concept of segregation had all been part of biological knowledge for some years. He made four deductions from his hypothesis and found that every one of them could be verified experimentally. He had been able to predict the expected results from crosses before they were made and later to find his predictions confirmed. In view of all this do you consider that his hypothesis was 'established beyond a reasonable doubt'?

Still another cross was made, but the results were most surprising. A *white*-eyed female was crossed with a *red*-eyed male. (The male was from a wild stock that had never been bred with the original stock that produced the *white*-eyed male.) All of the females derived from this cross had *red* eyes and all of the males had *white* eyes. One might have expected that all of the F_1 would have been *red*-eyed, since the male should have been of the genotype **WWX**. This was not the case, so Morgan assumed that all males he used were heterozygous for the *red* eye gene and had the genotype **WwX**:

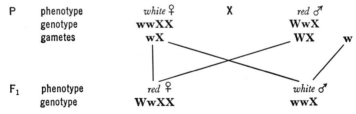

P	phenotype	*white* ♀	X	*red* ♂
	genotype	**wwXX**		**WwX**
	gametes	**wX**		**WX** **w**

F_1	phenotype	*red* ♀		*white* ♂
	genotype	**WwXX**		**wwX**

At this point we should pause and consider these questions:

a. Why are there only two classes of sperm, namely **WX** and **w**, formed by the heterozygous *red* males of the genotype **WwX**? Why is no **wX** class produced?

b. Why are all the wild *red*-eyed males heterozygous? Why does the **WWX** type not occur?

Morgan's hypothesis demands that *red*-eyed males are never homozygous and that they show unusual phenomena in sperm formation. If these basic conditions do not hold, then the hypothesis is either wrong or incomplete. Can you devise other hypotheses for explaining the data?

Explaining the Results: Second Hypothesis. It was not long—in fact, only one year—until Morgan devised a simpler hypothesis to account for the *white* eye case. *If one assumes that the gene for white eyes is actually part of the X chromosome, then the results of all the crosses correspond to what would be expected from the behavior of the X*

chromosome. There would then be no need for invoking subsidiary
assumptions, such as unusual types of meiosis in some males, or requiring all wild males to be heterozygous for eye color. Morgan's second
hypothesis has withstood every conceivable test, and there seems to be
no reasonable doubt of its correctness.

The symbolic representation for the new scheme will be different
from that given before. **W** will continue to mean *red* and **w** *white,* but
there will be no need to use **X**. If we assume that **W** and **w** are located
on the **X** chromosomes, '**W**' should be interpreted as an **X** chromosome
with the **W** gene. In the same manner, **w** will indicate an **X** chromosome with the **w** gene. The Y chromosome will be indicated by a **Y**,
since by this time it was realized that Drosophila is of the **XX** ♀ -**XY** ♂
sex chromosome type. The **Y** chromosomes contain no **W** or **w** genes.
(As later work was to show, the **Y** of *Drosophila melanogaster* is almost
entirely without genes.) The following, then, are the correct diagrammatic representations of the eye-color crosses through the F_2 generation:

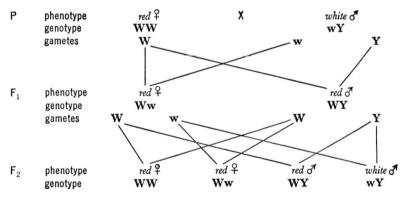

The reciprocal P generation cross would be as follows:

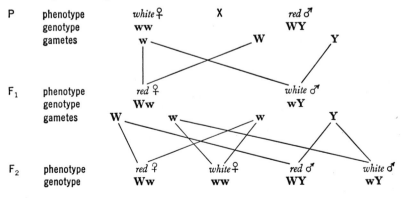

This new conceptual scheme explains the observed results of the genetic experiments, and it does not invoke any unknown phenomena.

Morgan soon discovered other genes which, from their mode of inheritance, he concluded were carried on the **X** chromosome. All genes that are on the **X** are said to be *sex linked* and they always show the type of inheritance just outlined for the *white* eye crosses. Those genes carried on any chromosome except the sex chromosomes are said to be *autosomal* genes.

AUTOSOMAL GENES

These experiments with *white*-eyed flies provided additional evidence supporting the hypothesis that chromosomes are the physical basis of inheritance. If some genes are parts of the **X** chromosome, their inheritance must reflect the behavior of the **X** during meiosis and fertilization. The *white* eye gene behaves as though it were part of the **X**. This can mean either that it is part of the **X** or that it is part of some unknown cell structure that behaves exactly like the **X** during meiosis, fertilization, and mitosis.

Crosses with Sex-linked and Autosomal Genes. Morgan and his associates discovered mutant genes by the dozens. Some were autosomal and some were sex linked. The inheritance of autosomal genes followed the usual Mendelian scheme. Sex-linked inheritance follows the scheme that has just been described.

The following example of a cross involving sex-linked genes and autosomal genes will bring out the difference between the two types. We already know that *white* eye is a sex-linked gene recessive to *red*. Our other characteristic in this cross is *vestigial* wing, which is an autosomal gene recessive to *long* wing.

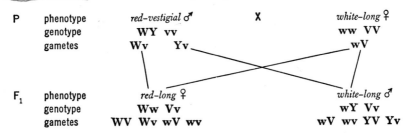

P	phenotype	*red–vestigial* ♂	**X**	*white–long* ♀
	genotype	**WY vv**		**ww VV**
	gametes	**Wv Yv**		**wV**

F₁	phenotype	*red–long* ♀		*white–long* ♂
	genotype	**Ww Vv**		**wY Vv**
	gametes	**WV Wv wV wv**		**wV wv YV Yv**

The F₂ which is shown in the checkerboard on the next page consists of the following:

♀ ⅜ *red-long;* ⅜ *white-long;* ⅛ *red-vestigial;* ⅛ *white-vestigial.*

♂ ⅜ *red-long;* ⅜ *white-long;* ⅛ *red-vestigial;* ⅛ *white-vestigial.*

		Sperm			
		wV	**wv**	**YV**	**Yv**
	WV	**Ww VV** *red-long* ♀	**Ww Vv** *red-long* ♀	**WY VV** *red-long* ♂	**WY Vv** *red-long* ♂
	Wv	**Ww Vv** *red-long* ♀	**Ww vv** *red-vestigial* ♀	**WY Vv** *red-long* ♂	**WY vv** *red-vestigial* ♂
Ova	**wV**	**ww VV** *white-long* ♀	**ww Vv** *white-long* ♀	**wY VV** *white-long* ♂	**wY Vv** *white-long* ♂
	wv	**ww Vv** *white-long* ♀	**ww vv** *white-vestigial* ♀	**wY Vv** *white-long* ♂	**wY vv** *white-vestigial* ♂

What is the ratio of *long* to *vestigial,* neglecting the eye-color genes? Is there any difference in the ratios of the autosomal genes between the F_2 males and females?

The Importance of Morgan's Work. These first experiments of Morgan are important in several ways. A new experimental animal was introduced to geneticists that was easy to raise in the laboratory and was a producer of large numbers of offspring. In addition, the crosses themselves added considerably to genetic theory in that they were the first well-analyzed cases of sex-linked inheritance. The fact that the genetic results exactly paralleled the behavior of the **X** chromosome was strong evidence that the gene responsible for *white* eyes is part of the **X** chromosome. At least many biologists believed the data to be highly suggestive. Now if it is established that one gene is part of a chromosome, it is a good working hypothesis that other genes are parts of chromosomes. One could even hold to the hypothesis that most or all genes are parts of chromosomes.

Scientific Methods. This early genetic work of Morgan is valuable in still another way. The experiments and the way in which they were reported are excellent examples of one of the most important procedures in experimental science, namely, the manner in which 'cause-effect' relations are discovered. A scientist is interested in the reason why things behave as they do. In this case, Morgan wondered what was the basic cause of the peculiar genetics of the *white* eye gene. He observed the effects and attempted to reconstruct the cause. This reconstruction, according to the philosophers who study scientific methods, takes place in well defined if not always explicitly stated steps, which are:

1. *Recognition of the problem.* In this instance the problem was to interpret the *white* eye case in genetic and cytological terms.

2. *Collection of facts pertaining to the problem.* The facts consisted of the data of the first cross together with all that Morgan knew of cytology and genetics.

3. *Formulation of a hypothesis.* From a consideration of all the particular facts, Morgan formulated a general statement, or *hypothesis,* that would explain the facts. This logical step from the particular to the general is known as *induction.* The hypothesis in this case was the symbolic scheme that explained the results of the cross in terms of chromosome behavior.

4. *Testing the hypothesis.* The correctness of a hypothesis is tested in this manner: First, we assume that the hypothesis is correct and then make certain deductions. These deductions can be tested to see if they are true or false. Morgan made four such deductions and found that the predicted results were always obtained. The more deductions that are verified, the more likely it is that the hypothesis is true.

The fate of Morgan's first hypothesis, which symbolized the *white* female as **wwXX** and the *white* male as **wwX**, should be a sobering example. It was tested by four deductions and found to be 'true.' For most scientists, this might be convincing. It did not, however, offer a convincing explanation of all the data. One had to assume that meiosis in the **WwX** males was of an unusual sort and that all *red*-eyed males are heterozygous. Morgan found, however, that a second hypothesis would explain the same data and in this case it was not necessary to introduce any qualifications, such as a special type of meiosis in males heterozygous for the eye-color gene or that all *red*-eyed males are heterozygous (**WwX**). The second hypothesis, which symbolized the *white* female as **ww** and the *white* male as **wY**, was simpler. When one has the choice of two hypotheses, one simple and one complex, one generally selects the first. This is the famous *Occam's razor,* which admonishes the scientist to explain his results in the simplest manner possible, and to introduce no unnecessary complexity. It must be realized that both of Morgan's hypotheses explained the data. Subsequent events have shown the first one to be false and the second, and simpler one, to be true.

This episode is an example of the self-correcting nature of scientific procedures. If deductions are made and tested, the truth or falsity of the hypothesis can be established. If the hypothesis fails to account for all the data, then it must be abandoned or modified. Morgan's first hypothesis accounted for most but not all of the experimental results. It was not necessary for him to abandon the hypothesis entirely, merely to modify it.

SUGGESTED READINGS

Morgan, T. H. 1910. 'Sex-limited inheritance in Drosophila.' *Science 32*:120–22.

10

Linkage and Crossing Over

If Sutton's hypothesis is correct, genes are parts of chromosomes. The inheritance of genes will therefore be identical with the inheritance of chromosomes. Stated in another way, all genetic theory must take into account the behavior of chromosomes, and all chromosome theory must take into account the results of genetic crosses.

The Prediction of Linkage. At the time Sutton proposed his hypothesis, he pointed out one situation in which the Mendelian laws could not apply, namely, those cases where two genes are carried on the same chromosome. Clearly, they would not obey Mendel's law of independent assortment. He foresaw that this problem would arise when more pairs of genes had been discovered than there are pairs of chromosomes in the species being studied.

Let us consider the problem as it applies to *Drosophila melanogaster*. The cells of this species have four pairs of chromosomes as the diploid number. Let us assume that each of the first four mutant genes discovered are located on a different chromosome pair. If this is the case, each of these four genes will show independent assortment. If we neglect the case of those carried on the **X**, we shall obtain a ratio of $9:3:3:1$ in the F_2 if the P generation cross was between an individual homozygous for both dominant genes and one recessive for both genes. What will happen when we discover the fifth pair of genes? Since there is no fifth pair of chromosomes, the fifth pair of genes must be located on a chromosome that already has one of the first four pairs of genes. When this situation arises, obviously the two pairs of alleles cannot act in an independent way during meiosis. They would be linked together in inheritance. This deduction is inevitable, if genes are on chromosomes.

The Discovery of Linked Genes. In 1906 Bateson and Punnett reported a cross involving two pairs of genes that did not show independent assortment. Their cross was with sweet peas, where *blue* flower color (**B**) is dominant over *red* (**b**) and *long* pollen grain (**L**) is dominant over *round* pollen (**1**). The scheme of the cross was this:

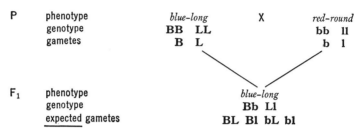

P	phenotype	*blue–long*	X	*red–round*
	genotype	BB LL		bb ll
	gametes	B L		b l

F₁	phenotype	*blue–long*
	genotype	Bb Ll
	expected gametes	BL Bl bL bl

If there is independent assortment in the Mendelian sense, we would expect the four classes of F₁ gametes shown in the diagram to be produced in equal numbers. Each type would account for 25 per cent of the total. The standard genetic way for finding gamete percentages is to cross the organism being tested with the pure recessive. This is called the *test cross.* For the F₁ heterozygous plant it would be this:

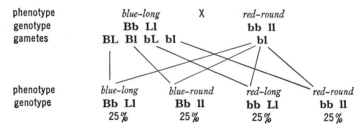

phenotype	*blue–long*	X	*red–round*
genotype	Bb Ll		bb ll
gametes	BL Bl bL bl		bl

phenotype	*blue–long*	*blue–round*	*red–long*	*red–round*
genotype	Bb Ll	Bb ll	bb Ll	bb ll
	25%	25%	25%	25%

In this test cross, the phenotype of the offspring would be a measure of the gamete frequency of the plant being tested. Thus, 25 per cent of the gametes would be **BL** and 25 per cent of the offspring would be *blue-long.* This would be true for all classes of gametes, since they would be combining with a gamete having both recessive genes. (These **bl** gametes, having only recessive genes, cannot alter the expression of genes in the gametes with which they combine).

When Bateson and Punnett made the cross, these were the results:

	EXPECTED	ACTUAL
blue-long	25%	43.7%
blue-round	25%	6.3%
red-long	25%	6.3%
red-round	25%	43.7%

Clearly these results do not conform to those expected from the Mendelian theory. Two points should be noticed.

1. The two most frequent phenotypes are those of the original parents (*blue-long* and *red-round*).
2. The percentages of the original parental types (*blue-long* and *red-round*) are the same and the percentages of the two recombination classes (*blue-round* and *red-long*) are the same.

An even more disturbing finding was that the F_2 ratios depended largely on the genotype of the P generation. On strict Mendelian principles a cross of *blue-long* × *red-round* should give the same F_2 as *blue-round* × *red-long*. This was not so with Bateson's sweet peas. We have already seen what the F_2 ratios were for the *blue-long* × *red-round* cross. The *blue-round* × *red-long* cross gave the following results (once again the results are compared with what would have been expected if Mendel's rules applied):

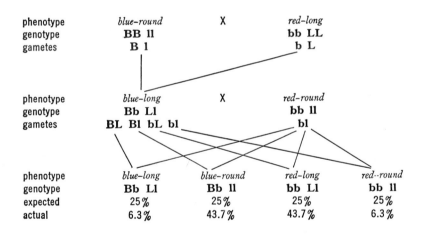

phenotype	*blue-round*	X	*red-long*
genotype	BB ll		bb LL
gametes	B l		b L

phenotype	*blue-long*	X	*red-round*
genotype	Bb Ll		bb ll
gametes	BL Bl bL bl		bl

phenotype	*blue-long*	*blue-round*	*red-long*	*red--round*
genotype	Bb Ll	Bb ll	bb Ll	bb ll
expected	25%	25%	25%	25%
actual	6.3%	43.7%	43.7%	6.3%

When these results are compared with those of the first cross, the percentages for each phenotype are found to be different, but again we notice a preponderance of the parental types. Both crosses suggest an orderly, though non-Mendelian, mechanism of inheritance. Some new principles must be involved.

Let us try to explain the results on the basis of Sutton's hypothesis. If we assume that the two genes, **B** and **L** are parts of the same chromosome, this will be the schematic representation of the cross:

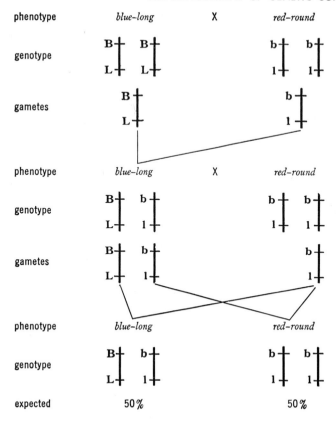

The percentages actually obtained can be compared with the ratios expected on the basis of Mendel's hypothesis and Sutton's hypothesis:

	blue-long	blue-round	red-long	red-round
ACTUAL	43.7	6.3	6.3	43.7
MENDEL'S	25	25	25	25
SUTTON'S	50	0	0	50

At first sight it may seem as though Sutton's explanation is the better since the *blue-long* and *red-round* classes are near the actual percentages. One difficulty, however, is fatal to Sutton's hypothesis. Both *blue-round* and *red-long* plants were obtained in the cross. Neither of these classes would be possible if both the **B** and **L** genes were part of the same chromosome. We may conclude that the results obtained in the cross cannot be understood on the basis of either Mendel's or Sutton's hypotheses, as these were originally stated.

Linkage and Crossing Over. The situation was eventually clarified by Morgan. New mutants were discovered in considerable numbers in his Drosophila experiments. He found that many crosses involving two pairs of genes gave the independent assortment expected in the Mendelian scheme. In other crosses, deviations of the sort found by Bateson and Punnett were encountered. Morgan was convinced of the correctness of Sutton's hypothesis that genes are parts of the chromosomes. He believed that the exceptions to independent assortment were due to the two different genes being on the same chromosome. But to explain the results satisfactorily, it was necessary to assume that under some circumstances genes could be transferred from one chromosome to another. Was there any cytological evidence for this?

A possible cytological basis for an exchange of genes was provided by Janssens. He described a type of behavior of chromosomes in meiosis that is now known to be of nearly universal occurrence in both animals and plants. It is called *crossing over* and it occurs during the tetrad stage (Fig. 10–1). Synapsis results in homologous chromosomes coming close together with their long axes parallel. Both chromosomes duplicate and a tetrad of four chromatids is formed. According to Janssens, there is considerable coiling of chromatids around one another at this time. In some cases two of the chromatids break at the corresponding place on each. The broken chomatids may rejoin in such a way that a section of one chromatid is now joined with a section of the other. As a result, 'new' chromatids are produced that are mosaics of segments of the original ones.

Janssens' theory of crossing over could provide the basis of gene

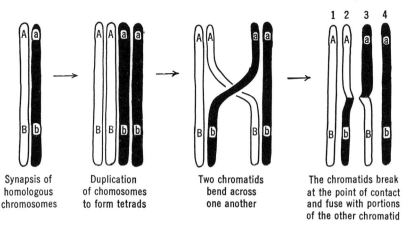

| Synapsis of homologous chromosomes | Duplication of chomosomes to form tetrads | Two chromatids bend across one another | The chromatids break at the point of contact and fuse with portions of the other chromatid |

10–1 Janssens' theory of crossing over.

transfer from one chromosome to another, and Morgan suspected that it did. This is what Morgan wrote:

In consequence, the original materials will, for short distances, be more likely to fall on the same side of the split, while remoter regions will be as likely to fall on the same side as the last, as on the opposite side. In consequence, we find linkage in certain characters, and little or no evidence at all of linkage in other characters; the difference depending on the linear distance apart of the chromosomal materials that represent the factors. Such an explanation will account for all of the many phenomena that I have observed and will explain equally, I think, the other cases so far described. The results are a simple mechanical result of the location of the materials in the chromosomes, and of the method of union of homologous chromosomes, and the proportions that result are not so much the expression of a numerical system as of the relative location of the factors in the chromosomes. Instead of random segregation in Mendel's sense we find 'associations of factors' that are located near together in the chromosomes. Cytology furnishes the mechanism that the experimental evidence demands.

The term *linkage* was introduced to refer to cases where different genes are located on the same chromosome. *Crossing over* was the term applying to the coiling, breaking, and rejoining of homologous chromosomes during meiosis.

The following is an example of inheritance of two linked genes: In Drosophila *gray* body color (**B**) is dominant to *black* body color (**b**). *Long* wing (**V**) is dominant to *vestigial* wing (**v**). The two pairs of genes are located on the same pair of autosomes.

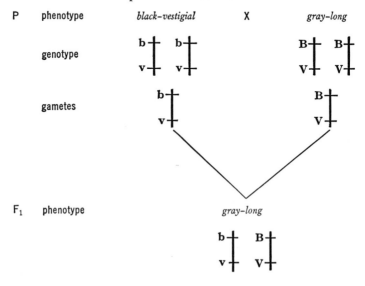

This F_1 individual will produce four types of gametes, two the result of crossing over and two non-crossovers. Since crossing over occurs in the tetrad stage we could diagram gamete formation as follows:

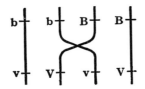

The middle pair of chromatids will break and recombine. The two meiotic divisions then occur and each resulting gamete will receive one chromosome. The four types of gametes will be as follows:

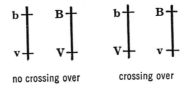

no crossing over   crossing over

These four gametes are not produced in equal frequency. Morgan pointed out that *the chance of a crossover occurring between two genes is a function of the distance between them.* The closer they are the smaller the chance of crossing over. If we mate an F_1 female of the above cross with a *black-vestigial* male, the results are as follows:

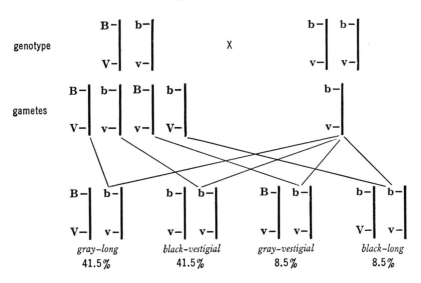

Among the offspring, 17 per cent are derived from gametes that carry chromosomes that had a crossover between the two genes being studied. The remainder, 83 per cent, are from non-crossover gametes.

The frequency of crossing over between any two genes is nearly constant. If we cross a *black-long* $\times$ *gray-vestigial* fly the F_1 would be *gray-long* and heterozygous for both genes. In these two respects the F_1 of this cross will be identical with that in the one previously described. In this second cross, however, one of the P generation chromosomes will be **bV** and the other **Bv**. If a female of this constitution is crossed with a *black-vestigial* male the offspring will be as follows:

> 41.5% *black-long*
> 41.5% *gray-vestigial*
> 8.5% *gray-long*
> 8.5% *black-vestigial*

Compare these percentages with the previous cross and be sure you understand the reason for the difference.

Subsequent events have shown that Morgan's explanation satisfactorily accounts for the inheritance of genes located on the same chromosome. Many details were added, such as the occurrence of double or triple crossovers, and the absence of crossing over in Drosophila males.

The analysis of these experiments is an example of the mutual checking that the combined genetic and cytological approach permits. As we have mentioned before, inheritance must be explained in both fields. If two genes are in the same chromosome they will be linked in inheritance. If they are not completely linked there must be a chromosomal basis for the recombination. A chromosomal basis is to be found in the phenomenon of crossing over.

Linkage Groups in Drosophila. Another interesting parallel between genetics and cytology was soon apparent. One deduction we could make from Sutton's hypothesis is this: If genes are on chromosomes and all chromosomes have genes, then the number of groups of linked genes would obviously correspond to the number of pairs of homologous chromosomes. This deduction was verified. By 1915 Morgan and his associates had studied more than 100 mutant genes. When these were tested, *they were found to comprise four linkage groups. The number*

10-2   A total of 85 genes of *Drosophila melanogaster* were reported on in 1915. These fell into 4 linkage groups. Cytological investigations showed that this species has 4 pairs of chromosomes. This parallelism between the number of chromosomes and the number of linkage groups suggested that the genes were situated on the chromosomes (T. H. Morgan, 'The Constitution of the Hereditary Material,' *Proc. Amer. Phil. Soc.* 54:143–53. 1915).

GROUP I

Name	Region Affected
Abnormal	Abdomen
Bar	Eye
Bifid	Venation
Bow	Wing
Cherry	Eye color
Chrome	Body color
Cleft	Venation
Club	Wing
Depressed	Wing
Dotted	Thorax
Eosin	Eye color
Facet	Ommatidia
Forked	Spines
Furrowed	Eye
Fused	Venation
Green	Body color
Jaunty	Wing
Lemon	Body color
Lethals, 13	Die
Miniature	Wing
Notch	Venation
Reduplicated	Eye color
Ruby	Legs
Rudimentary	Wings
Sable	Body color
Shifted	Venation
Short	Wing
Skee	Wing
Spoon	Wing
Spot	Body color
Tan	Antenna
Truncate	Wing
Vermilion	Eye color
White	Eye color
Yellow	Body color

GROUP II

Name	Region Affected
Antlered	Wing
Apterous	Wing
Arc	Wing
Balloon	Venation
Black	Body color
Blistered	Wing
Comma	Thorax mark
Confluent	Venation
Cream II	Eye color
Curved	Wing
Dachs	Legs
Extra vein	Venation
Fringed	Wing
Jaunty	Wing
Limited	Abdominal band
Little crossover	II chromosome
Morula	Ommatidia
Olive	Body color
Plexus	Venation
Purple	Eye color
Speck	Thorax mark
Strap	Wing
Streak	Pattern
Trefoil	Pattern
Truncate	Wing
Vestigial	Wing

GROUP III

Name	Region Affected
Band	Pattern
Beaded	Wing
Cream III	Eye color
Deformed	Eye
Dwarf	Size of body
Ebony	Body color
Giant	Size of body
Kidney	Eye
Low crossingover	III chromosome
Maroon	Eye color
Peach	Eye color
Pink	Eye color
Rough	Eye
Safranin	Eye color
Sepia	Eye color
Sooty	Body color
Spineless	Spines
Spread	Wing
Trident	Pattern
Truncate intensf.	Wing
Whitehead	Pattern
White ocelli	Simple eye

GROUP IV

Name	Region Affected
Bent	Wing
Eyeless	Eye

of *chromosome pairs in Drosophila is also four.* The partial list that Morgan published at that time is given in Fig. 10–2.

The evidence was becoming almost overwhelming that Sutton's hypothesis was correct, though it was necessary to modify it to take crossing over into account.

In Fig. 10–2, two sets of data are given, namely, the genetic list of linked genes and the cytological picture of the chromosomes. As we have seen, there is good reason to believe that a relation exists between the two sets of data.

If we accept the hypothesis that the linkage groups correspond to the pairs of homologous chromosomes, how could we determine which linkage group corresponds to each of the four pairs of chromosomes? It will be worthwhile for you to consider this problem.

Relation of Genes to Characteristics. The data in Fig. 10–2 showing the linkage groups of Drosophila are instructive in another connection. Notice that many different genes affect the same character: 13 influence eye color and 33 modify the wings in some manner. The question arises, what determines the normal red eye color? The answer is that the wild type alleles of all of these 13 eye color genes, together with many undiscovered in 1915 when Morgan published his list, act together to produce the wild type red eye color. If an individual is homozygous for the mutant allele of any one of these genes, then the eye is not red but some other color such as white, sepia, or peach. We should think of the normal red eye color as the end product of a series of gene actions. If any of these actions is altered, the eye color will be something different.

Figure 10–2 is misleading in one respect. Each mutant gene appears to have a single effect. It usually does have a single *main* effect, but most of the genes that have been studied intensively are found to have many different effects. Thus the *white* eye color gene in Drosophila is responsible not only for the absence of color in the compound eyes but also for the absence of color in the simple eyes and in some of the internal organs as well. It is called an eye color gene simply because the most obvious effect of the gene is on the color of the compound eyes. Genes that affect more than one structure are said to be *pleiotropic.* Some geneticists now believe that every gene has some effect on every character, i.e. every character is affected, to some extent, by every gene.

THE CYTOLOGICAL PROOF OF CROSSING OVER

With the data so far given, the concept of crossing over might be regarded as a good working hypothesis and nothing more. In a sense the hypothesis was invented by geneticists to account for the results of genetic crosses. The following is a brief recapitulation of what occurred: The absence of recombination of genes in some crosses was interpreted to indicate that the genes in question were parts of the same chromosome. This explanation did not account for all the data, however, since in a definite percentage of the individuals recombination did occur. Now if the genes in question are parts of chromosomes, these recombinations among genes of the same linkage group could only be explained on the basis of some exchange of genes between homologous chromosomes.

After the Morgan group had analyzed the situation to this extent they sought some possible cytological basis for the postulated interchange of genes between chromosomes. It was then that they came across the work of Janssens, who had described a phenomenon that *might be interpreted* in terms of the breaking and rejoining of chromatids during the tetrad stage of meiosis. It should be emphasized that Janssens did not actually observe the breaking and rejoining of chromosomes and no one has to this day. The difficulty is this. Crossing over is assumed to occur between homologous chromatids. Since they are homologous, they are of identical appearance when viewed under the microscope. Furthermore, the act of crossing over is assumed to occur when the four chromatids of the tetrad are tightly coiled around one another. Crossing over cannot be seen in living cells, and in fixed and stained cells there is no direct way of telling whether a chromatid has exchanged portions with another chromatid or not.

Figure 10–1 is a diagram of crossing over. The two homologous chromosomes undergoing synapsis are drawn differently, but it must be remembered that in living or in fixed and stained material they would be of identical appearance. In the four chromatids shown after crossing over there is no difficulty in distinguishing the chromatids that have crossed over and those which have not, since the strands have been shaded differently by the artist. Once again this is impossible to observe in the actual material.

One could obtain critical cytological evidence for crossing over if there was some visible or detectable difference between the members of a homologous pair of chromosomes. Such evidence was not available for animals until the work of Stern in 1931. This was nearly 20 years later than the time Morgan's group postulated the existence of crossing over. We shall consider Stern's work out of turn, so to speak, but by so doing we can complete the topic under consideration.

By the time Stern began his work, Drosophila geneticists had a large variety of strains with different types of chromosome abnormalities. He was able to obtain a female that had the necessary chromosomal and genetic characteristics to demonstrate whether or not crossing over involves a transfer of material from one chromosome to another.

The female used had two structurally and genetically different **X** chromosomes (Fig. 10–3). One of the **X** chromosomes was broken into two portions: one portion behaved as an independent chromosome and the other was attached to one of the tiny fourth chromosomes (Fig. 10–2 shows the chromosomes of a normal ♀; the fourth chromosomes are the pair of dot-shaped structures). The other **X** of this female was unusual in that a piece of a **Y** chromosome was attached to it. These structural differences were so great that they could be seen easily in fixed and stained nuclei.

The two **X** chromosomes, in addition to being structurally different, were also genetically different. The divided **X** had in one portion of it the recessive gene *carnation* (**c**), which when homozygous produces a dark ruby eye color, and the dominant gene *bar* (**B**), which reduces the eye to a narrow band. The other **X**, which had the piece of the **Y** attached to it, contained the wild type alleles, **C** and **b**, which when homozygous result in *red* eyes of *normal* shape.

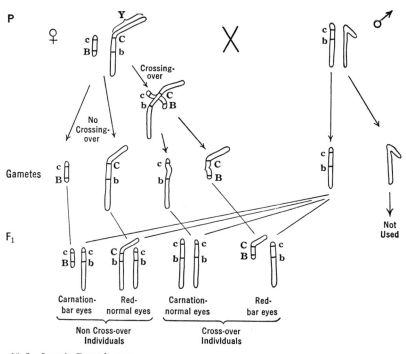

10–3   Stern's Experiment

The essential point about this female is that she had two **X** chromosomes that could be distinguished from one another on both cytological and genetic grounds.

This female was crossed to a male fly carrying the genes for *carnation* color (c) and *normal* eye shape (b). The ova of the female would be of two types if no crossing over occurred: one type of ova would contain the short **X** chromosome with the genes c and **B**; the other type would have the **C** and b genes on the **X** that had the piece of **Y** chromosome attached to it. If crossing over occurred between the two marker genes, two other types of gametes would be produced. One of these crossover types would have the c and b genes on an **X** chromosome of normal size; the other crossover type would have the **C** and **B** genes on a short **X** chromosome to which the piece of **Y** chromosome was joined.

Stern studied only the female offspring of the cross. One can determine from Fig. 10–3 that four types of daughters are to be expected, showing all combinations of the phenotypic characters. These flies should also have four different chromosome configurations, and if the theory is correct it should be possible to predict the chromosome configuration for each phenotypic class. Thus, the flies that give genetic evidence of coming from crossover gametes will be either *carnation-normal* or *red-bar*. The *carnation-normal* flies, alone among the offspring, should have two normal-shaped **X** chromosomes. The *red-bar* flies, again alone among the offspring, should have one short **X** with a piece of **Y** chromosome attached and an **X** of normal proportions.

Stern studied the cytology of his flies and saw that the phenotype corresponded to the expected chromosomal configuration. This was a brilliant demonstration of the hypothesis that chromosomal material can be interchanged between homologous chromosomes.

SUGGESTED READINGS

Morgan, T. H. 1919. *The Physical Basis of Heredity*. Lippincott.

Morgan, T. H., A. H. Sturtevant, H. J. Muller, and C. B. Bridges. 1915 and 1919. *The Mechanism of Mendelian Heredity*. Holt.

Stern, C. 1931. 'Zytologisch-genetische Untersuchungen als Beweise für die Morgansche Theorie des Faktorenaustauschs. *Biologisches Zentralblatt. 51:* 547–87.

Wilson, E. B. 1928. *The Cell in Development and Heredity*. Macmillan. Section III of Chapter 12 discusses crossing over.

11

Mapping the Genes

Morgan and his fellow workers made numerous crosses involving linked genes. It was found that, in successive experiments, the amount of crossing over between two particular genes was always the same. Depending on the genes used, the amount of crossing over might be less than 1 per cent or nearly 50 per cent. Morgan suggested that the different values were the result of the relative positions of the genes on the chromosome: If the amount of crossing over between hypothetical genes **A** and **B** was small and between genes **C** and **D** large, one would predict that **A** would be closer to **B** than **C** would be to **D**.

The development of this concept, that linkage data could be used to map the relative positions of the genes on the chromosomes, was attempted in 1913 by Sturtevant, a student of Morgan. He made crosses involving five genes carried on the **X** chromosome, namely, *yellow* body (**y**), *white* eyes (**w**), *vermilion* eyes (**v**), *miniature* wings (**m**), and *rudimentary* wings (**r**). From the data obtained, he constructed a genetic map showing the 'positions' of these genes on the **X** chromosome. This was his basic assumption: 'It would seem . . . that the proportion of "crossovers" could be used as an index of the distance between any two factors. Then by determining the distances (in the above sense) between A and B and between B and C, one should be able to predict [the amount of crossing over in the interval] AC. For, if proportion of crossovers really represents distance, AC must be approximately, either AB plus BC, or AB minus BC, and not any intermediate value.'

The percentage of crossovers between y and v was found to be 32.2 and between y and **m** 35.5. On the basis of the hypothesis we would expect **v** to be closer to **y** than **m** to **y**. What can we conclude about the relative positions of **m** and **v**? According to Sturtevant this should be 67.7 (35.5 + 32.2) or 3.2 (35.5 − 32.2). The reason for this is as follows: The chromosome is known to be a very long and very thin structure. We can consider it as having length only, and can represent it as a long line. On this line we shall put y and v, as follows:

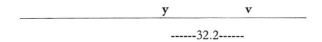

Now **m** can be either to the right or to the left of y, as shown here:

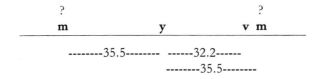

If **v** and **m** are on the same side of y, we would expect the amount of crossing over between **v** and **m** to be 3.2 per cent. If they are on opposite sides of y the value should be 67.7 per cent. When Sturtevant measured the amount it was found to be 3 per cent, which indicated that v and m were on the same side of y. This close correspondence between the actual and expected result was strong support for his theory.

In this manner, the relative positions of the five genes were determined and a genetic map constructed. The y gene was taken as the reference point and the distances measured from it. This was the result.

y	w		v	m		r
0.0	1.0		30.7	33.7		57.6

It was found that the most reliable values were obtained when crossover values for adjacent genes were used. This was due to the occurrence of double crossover, which introduced an error into the results. Let us consider the three genes y, w, and **m** (Fig. B11–1). Sturtevant raised 10,495 flies to test the linkage relations. He found that in 6,972 flies there was no crossing over between the three genes. In 3,454, crossovers occurred between **w** and **m**; in 60, crossovers between y and **w** were detected; and in nine a double crossover occurred. That is, there was one crossover between y and **w** and another between **w** and **m**. This would result in y and **m** being on the same chromosome as they were before the two crossovers occurred.

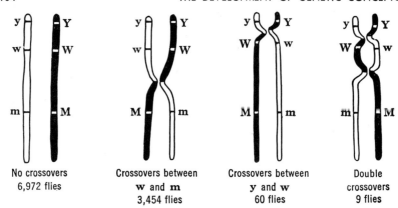

No crossovers	Crossovers between	Crossovers between	Double
6,972 flies	**w** and **m**	**y** and **w**	crossovers
	3,454 flies	60 flies	9 flies

11-1   Sturtevant's Experiment

In the case of the double crossovers it should be noted that three genes are always necessary to detect the event. Thus, if only the genes **y** and **m** were used, any double crossover between them would be undetected since **y** and **m** would still be together after the chromatids had broken and rejoined. When genes are far apart, double crossovers are likely. If they are not detected they will introduce an error in the positions assigned to the genes, for the data would indicate that the genes are closer to one another than they really are. For this reason Sturtevant suggested that chromosome maps be based on crossover values of genes close to one another and not those far apart.

Now the question arises, What is the relation of the chromosome map to the position of these genes on the chromosome? Sturtevant has this to say:

Of course, there is no knowing whether or not these distances as drawn represent the actual relative spatial distances apart of the factors. Thus, the distance **wv** may in reality be shorter than the distance **yw**, but what we do know is that a break is far more likely to come between **w** and **v** than between **y** and **w**. Hence, either **wv** is a long space, or else it is for some reason a weak one. The point I wish to make here is that we have no means of knowing that the chromosomes are of uniform strength, and if there are strong or weak places, then that will prevent our diagram from representing actual relative distances —but, I think, will not detract from its value as a diagram.

Sturtevant's chromosome map was a graphic way of expressing linkage data. Once constructed, these maps proved useful in predicting the results of untried crosses. The most important induction from the data is that the genes are arranged in a linear order on the chromosomes, analogous to the sequence of beads on a string. The position occupied by a gene is its *locus*.

SUGGESTED READINGS

Sturtevant, A. H. 1913. 'The Linear Arrangement of Six Sex-linked Factors in Drosophila, as Shown by Their Mode of Association.' *Journal of Experimental Zoology* 14:43–59.

12

The 'Final Proof' That Genes Are Located on Chromosomes. Sex Determination Continued

BRIDGES' EXPERIMENTS WITH NON-DISJUNCTION FLIES

Beginning in 1884 with Hertwig and others, we have seen that *some* biologists thought the evidence indicated that the hereditary factors were parts of the chromosomes. It is probable that a minority held this view prior to 1910. After 1910 the Drosophila data collected by Morgan and his associates made it increasingly probable that genes are parts of chromosomes, and more and more biologists came to accept this view. Geneticists generally credit the work of Bridges, published in 1914 and 1916, as being the 'final proof' that the genes are located in the chromosomes. The material that will now be covered should be studied more to learn the type of evidence constituting a 'final proof,' and less for the genetic details.

Bridges' experiments dealt with the inheritance of sex-linked genes in Drosophila. The hypothesis that he sought to prove was that 'sex-linked genes are located on the sex chromosomes.'

Normal Inheritance of Sex Chromosomes. In order to understand Bridges' experiments, it is necessary to have clearly in mind the normal inheritance of sex chromosomes. The X chromosome of a male is transmitted only to his daughters and his Y only to his sons. The X chromosomes of a female are transmitted to both sons and daughters. Looked at from the point of view of the offspring, a daughter receives one X from her father and one from her mother. The sons receive an X from the mother and a Y from the father. The following diagram depicts this. In it, the sex chromosomes of the female are indicated in large letters and those of the male in small letters.

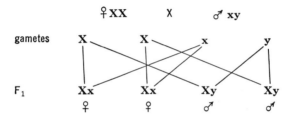

Inheritance in Non-disjunction Females. Bridges noticed that in some strains of Drosophila the inheritance of sex-linked genes was most unusual. Thus, in a cross between a *white*-eyed ♀ and a *red*-eyed ♂, some *white*-eyed daughters and *red*-eyed sons were obtained. These daughters had inherited their sex-linked genes solely from the mother, and the sons had inherited their sex-linked genes solely from the father. This would be impossible if (a) the sex-linked genes were located on the **X** chromosome, and (b) the sex chromosomes were inherited as shown in the diagram.

Bridges realized that the unexpected breeding results could be explained on the assumption that the female parent giving the unexpected offspring had two **X** chromosomes plus one **Y**. We could designate her **XXY**, in contrast with a normal female, which is **XX**. During meiosis a normal female produces only one class of ova so far as the sex chromosomes are concerned, namely, those with a single **X**. An **XXY** female would produce four types of ova during meiosis. These would be **XY, X, XX,** and **Y**. There was no way of predicting the frequency of each type of gamete, but we will anticipate the breeding results where it was found that the proportions were: 46 per cent **XY**; 46 per cent **X**; 4 per cent **XX**; 4 per cent **Y**.

Females of the **XXY** type were called *non-disjunction* females. The term refers to the fact that in some of the ova produced by these females there is no separation, or disjunction, of the two **X** chromosomes. In a normal female there is regularly a disjunction of the two **X** chromosomes with the result that a single **X** is present in each ovum.

The cross of a *white*-eyed non-disjunction female to a normal *red*-eyed male according to Bridges' hypothesis would be as shown in Fig. 12–1.

It must have taken considerable courage to postulate such a seemingly preposterous hypothesis, although some such hypothesis was necessary to explain the results, if one were to continue to hold the belief that genes are located on chromosomes. The hypothesis could be verified, however, since certain deductions were possible, and these deduc-

P

Non-Disjunctional
White Eye ♀
𝕏𝕏𝕐   X

Normal Red
Eye ♂
XY

Gametes

𝕏𝕐 (46%); 𝕏 (46%)   **X** (50%)
𝕏𝕏 (4%); 𝕐 (4%)   **Y** (50%)

F₁

	𝕏𝕐 (46%)	𝕏 (46%)	𝕏𝕏 (4%)	𝕐 (4%)
X 50%	**1** 𝕏**X**𝕐 23% Red Eye ♀. Would show non-disjunctional behavior if crossed	**2** 𝕏**X** 23% Red Eye ♀. Normal chromosome behavior	**3** 𝕏𝕏**X** 2% Triploid X ♀. Usually dies	**4** **X**𝕐 2% Red Eye ♂. The X has come from the father and the Y from the mother. This is the reverse of the normal situation.
Y 50%	**5** 𝕏𝕐**Y** 23% White Eye ♂. With extra Y chromosome	**6** 𝕏**Y** 23% White Eye ♂. With normal chromosome behavior	**7** 𝕏𝕏**Y** 2% White Eye ♀. Would show non-disjunctional behavior if crossed	**8** 𝕐**Y** 2% Dies

12–1 **Bridges' Experiment**

tions were open to test by observation and experiment. If the deductions were verified, the hypothesis would be established as 'true.' These were the main deductions:

1. If the hypothesis is true, we would expect 50 per cent of the daughters (classes 1 and 7 of Fig. 12–1) to be non-disjunction females. Breeding experiments showed this to be the case.

2. If the hypothesis is true, we would expect the exceptional ♂ (class 4) not to transmit the power of producing exceptions in later generations. It should behave like a normal male. Breeding experiments showed this to be true.

3. If the hypothesis is true, we would expect 46 per cent of the males to be **XYY**. These would produce sperm of four genotypes, namely, **X, YY, XY,** and **Y**. If a male of this type were crossed to a normal female, there should be no exceptional offspring (i.e. males inheriting their sex-linked characteristics only from the father, and females inheriting theirs only from the mother). However, every **XY** sperm that

entered a normal **X**-containing egg would produce an **XXY** daughter, who should be a non-disjunctional female. Breeding experiments confirmed all these predictions.

4. If the hypothesis is true, we would expect that 50 per cent of the daughters (classes 1 and 7) would have two **X** chromosomes and one **Y**. It should be possible to verify this deduction by cytological examination of the F_1 females. Bridges did this and found that approximately half of the daughters that he examined had two **X** chromosomes plus one **Y** (Fig. 12–2). The other half had two **X** chromosomes only. This was the crucial test of the hypothesis, since the test was of a very different sort, namely, cytological. (We can ignore the rare **XXX** females.)

Bridges' conclusion was as follows: '. . . there can be no doubt that the complete parallelism between the unique behavior of the chromosomes and the behavior of the sex-linked genes and sex in this case means that the sex-linked genes are located in and borne by the X-chromosomes.'

After this work of Bridges', nearly all geneticists believed that the sex-linked genes of Drosophila were parts of the **X** chromosome. It seemed equally probable that the autosomal genes were likewise parts of chromosomes, though this was not proved by his experiments. A further extension was made to other species, and the conclusion was reached that the genes of all organisms are parts of chromosomes. This extrapolation from the data was done because it appeared that inheritance was the same in all organisms being studied.

It might be of interest to inquire about the nature of this 'final proof,' in 1914, that genes are located on chromosomes. It must be apparent that it is the same type of evidence that had been offered

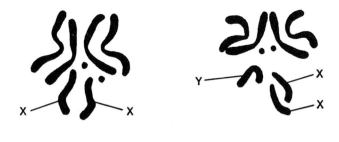

a   b

12–2   Bridges' drawings of the chromosomes of the females in his non-disjunction experiment. Approximately half of the females (class 2) had the normal chromosome complement as shown in *a*. The remaining females (classes 1 and 7) have two X chromosomes and a Y as shown in *b* (C. Bridges, 'Non-disjunction as proof of the chromosome theory of heredity,' *Genetics* 1:1–51; 107–63. 1916).

ever since 1902. Sutton pointed out the parallel behavior of chromosomes in meiosis and fertilization with the behavior of Mendelian factors. This evidence probably convinced a few that Mendelian factors were on the chromosomes. Morgan's analysis of the *white* eye case in Drosophila showed that inheritance of the gene was an exact parallel to the inheritance of the **X** chromosome. This study convinced a large segment of biologists. The discovery that the number of linkage groups is the same as the number of chromosome pairs was further support for the theory. These and many other experiments showed that either the genes were parts of chromosomes or the genes were parts of structures that behaved precisely like the chromosomes.

Bridges' evidence was of the same type, though it differed in degree. The inheritance of eye color in his non-disjunction experiments was completely different from any other type of inheritance. If he assumed genes were carried on chromosomes, then he had to postulate some most unusual chromosome phenomena. Cytological studies verified the predictions made from the genetic data. There could no longer be a 'reasonable doubt' that the genes were on chromosomes. More elaborate evidence was still to come, but for most biologists this evidence of Bridges was sufficient.

There were a few cases of inheritance that seemed to follow a different pattern. These were grouped under the term *cytoplasmic inheritance,* since it seemed that some non-nuclear factor was responsible. Over the course of the years, many cases that seemed to be due to cytoplasmic inheritance were found to be misinterpretations of the data. A few instances of cytoplasmic inheritance are well established. For the other thousands of analyzed cases, there is no doubt that the genes are parts of chromosomes. The chromosomes form the physical basis for 99.9+ per cent of inheritance.

THE CHROMOSOME BALANCE THEORY OF SEX DETERMINATION

The problem of sex determination as it was understood in the first decade of the twentieth century was discussed in Chapter B-7. In that chapter we learned that there was a constant relation between the sex of an organism and the type of chromosomes which it possessed. The cells of *Drosophila melanogaster* females contained three pairs of autosomes and two **X** chromosomes. In the cells of males of this species there were three pairs of autosomes and one **X** and one **Y** chromosome.

Genetic work with *Drosophila melanogaster* revealed that the **Y** chromosome contained very few genes, although it is essential for fertility in males. The **Y** was looked upon as a nearly inert chromosome genetically; a view that was strengthened by the discovery that in many animal species the males have a single **X** and no other sex chromosome. These data led to the concept that in species with a **XX** ♀ -**XY** ♂ sex chromosome constitution, the presence of a single **X** determined that the individual be a male and a pair of **X** chromosomes determined that the individual be a female.

This hypothesis was strengthened further by some remarkable observations on gynandromorphs in Drosophila. Gynandromorphs are individuals in which part of the body has the morphological features of a male and the other part has the morphological features of a female. Morgan and Bridges discovered some gynandromorphs that were female on one side and male on the other. Analysis showed that these individuals began development as females. Due to some developmental accident in the early embryo, one **X** chromosome was lost from the cells that were to form one-half of the body. As a result the cells of one side of the body remained normal and contained the three pairs of autosomes and two **X** chromosomes. This side had the external structure typical of females and internally an ovary might be present. The cells of the other side of the body where one **X** was lost contained three pairs of autosomes and a single **X** chromosome. This side had the external structure typical for males and internally a testis might be present.

These observations greatly strengthened the concept that a fly was a male or female depending on the number of **X** chromosomes present in its cells. This concept seemed to be an adequate explanation of the data and geneticists were willing to accept it so long as further tests were not possible.

The work of Bridges on non-disjunction, which we have just discussed, and of others on similar problems, showed that a remarkable amount of chromosome juggling was possible in Drosophila. As a consequence it became feasible to further test the hypothesis that one dose of **X** resulted in a male and two doses of **X** resulted in a female.

This concept was soon found to be inadequate to explain all of the data. Some Drosophila were obtained that were triploid, that is, there were three **X** chromosomes and three members of each autosome type in every cell. These individuals were females.

Bridges obtained flies with various combinations of autosomes and sex chromosomes (Fig. 12–3) and saw that sex was due to a balance between the number of **X** chromosomes and the number of autosomes. In a normal ♀ there are two **X** chromosomes and two haploid sets of autosomes (a haploid set of autosomes consists of one of each of the

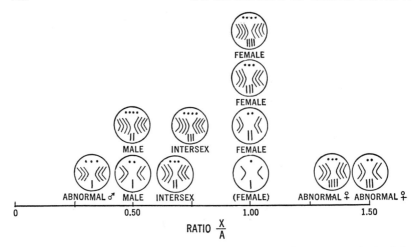

12-3   The various combinations of X chromosomes and autosomes obtained by Bridges and others. (The lowest circle at ratio 1.00 is a haploid female. No such fly has been obtained but some diploid flies have been observed which have haploid areas in their bodies. If these areas include sex structures, they are of the female type.)

three different kinds of autosomes). We could express this as $2X/2A = 1.0 = ♀$. A normal male would be $1X/2A = 0.5 = ♂$. The triploid female would be $3X/3A = 1.0 = ♀$.

When a triploid female is crossed with a diploid male, some of the offspring have two X chromosomes and three members of each autosome. We could write these as $2X/3A = 0.67$. Now this ratio is intermediate between the value for normal males (0.5) and that for normal females (1.0) and it was observed that these individuals were intermediate in appearance between males and females. They are described by Bridges as follows:

The 'intersexes,' which were easily distinguished from males and from females, were large-bodied, coarse-bristled flies with large roughish eyes and scolloped wing-margins. Sex-combs (a male character) were present on the tarsi of the fore-legs. The abdomen was intermediate between male and female in most characteristics. The external genitalia were preponderantly female. The gonads were typically rudimentary ovaries; and spermathecae were present. Not infrequently one gonad was an ovary and the other a testis; or the same gonad might be mainly ovary with a testis budding from its side.

The intersexes were sterile.

It was possible to obtain individuals with ratios of X chromosomes to autosomes that were below the normal male value or above the normal female value. Some individuals had three X chromosomes and two sets of autosomes. The ratio for these would be $3X/2A = 1.5$. This ratio is higher than that of a normal female, and the resulting imbalance in **X** chromosomes and autosomes produces a sterile and somewhat abnormal female. Bridges was able to obtain a value below the 0.5 ratio of normal males in individuals with a single **X** and three autosome sets. These would be $X/3A = 0.33$. Such flies were structurally abnormal and sterile males.

These and many other combinations of chromosomes were obtained by Bridges and others. The results formed a consistent pattern, there being a relation between the ratio of **X** chromosomes to autosome sets and the sex characteristics of the flies.

RATIO X/A	MORPHOLOGICAL TYPE
0.33	abnormal male
0.50	male
0.67	intersex
0.75	intersex
1.00	female
1.33	abnormal female
1.50	abnormal female

The significance of these ratios is to be found in the differential effectiveness of genes on the autosomes and on the **X** chromosomes. The net effect of the autosomal genes is a male-forming tendency. The net effect of the **X** chromosome genes is a female-forming tendency. In a normal male the genes of the two autosome sets overbalance the genes· of the single **X** to produce the male. In the normal female the **X** chromosome genes are in a double dose and this is sufficient to produce the female body type.

SUGGESTED READINGS

Bridges, C. 1914. 'Direct Proof through Non-disjunction that the Sex-linked Genes of Drosophila Are Borne by the X-Chromosome.' *Science 40*:107–9.

Bridges, C. 1916. 'Non-disjunction as Proof of the Chromosome Theory of Heredity.' *Genetics 1*:1–52, 107–63.

Bridges, C. 1921. 'Triploid Intersexes in Drosophila melanogaster.' *Science 54*:252–4.

Bridges, C. 1939. 'Cytological and Genetic Basis of Sex.' In *Sex and Internal Secretions* edited by E. Allen. Williams and Wilkins.

13

Multiple Alleles and Human Blood Type Genes

In all the cases considered so far, a gene has existed in only two states. It might be involved in the production of *red* or *white* eyes, *long* or *vestigial* wings, *round* or *wrinkled* peas. The *white* eye gene appeared in a stock of *red*-eyed flies. Since the stock had been under observation for many generations, it is reasonable to assume that in one **X** chromosome the gene, which in the normal condition produced *red* eyes, changed in such a way as to produce *white* eyes. One question this suggests is: If the original *red* eye gene can change to *white*, might it not change in another way to produce a still different result? Continued observations answered this question in the affirmative. A mutant known as *eosin* was discovered. Its phenotypic expression was a diluted red eye color. In its linkage relations and crossover behavior, it was found to occupy the same place on the **X** chromosome as the gene for *white*. In crosses involving the *red, white,* and *eosin* genes, it was never possible to have more than two of the three in the same female. All of the data were consistent with the hypothesis that *red, white,* and *eosin* were different states of the same gene. Phenomena of this sort are known as *multiple alleles*. At the present there are several dozen known alleles at the *white* locus. In other words, this gene is known to exist in several dozen states that can be distinguished.

The A, B, O Blood Type Genes. A well-known case of multiple alleles in man is the blood type series. The blood types, *A, B, AB,* and *O* are determined by the interaction of three autosomal alleles, **A, B,** and **o.** A and **B** are dominant over **o.** When **A** and **B** are present in the same individual neither gene is dominant and the individual is type *AB.* The phenotypes and possible genotypes are as follows:

PHENOTYPE	GENOTYPE
Type A	AA or Ao
Type B	BB or Bo
Type AB	AB
Type O	oo

The inheritance of these alleles follows the usual Mendelian scheme. A cross of a heterozygous type A with a heterozygous type B would be as follows:

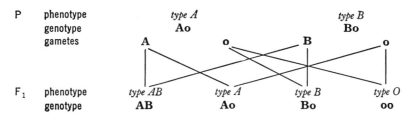

P	phenotype		*type A*				*type B*	
	genotype		Ao				Bo	
	gametes	A		o	B			o

F_1	phenotype	*type AB*	*type A*	*type B*	*type O*
	genotype	AB	Ao	Bo	oo

Digression on the Importance of Blood Types in Transfusions. Some years before the genetic basis of blood groups had been worked out, Landsteiner found that human blood could be classified into the four types just described. The types proved to be of great importance in connection with blood transfusions. In some cases death resulted when the donor and recipient were of different types. Experimentation and observation revealed that this incompatibility was due to the interaction of *antigens* on the surface of the red blood corpuscles with *antibodies* in the plasma. These blood type antigens and antibodies are proteins. There are two types of antigens, A and B, and two types of antibodies, α and β. The distribution of these substances is as follows:

BLOOD TYPE	ANTIGEN IN CORPUSCLE	ANTIBODY IN PLASMA
A	A	β
B	B	α
AB	A and B	none
O	none	α and β

The corpuscles are agglutinated (clumped) if those containing A antigen come in contact with α antibody, or if those containing B antigen come in contact with β antibody. The important factor is the type of corpuscle introduced in a transfusion; the introduced plasma has little or no effect on the recipient's corpuscles. Any interaction, therefore, will be between the donor's antigens on the corpuscles and the recipient's antibodies in the plasma, resulting in clumping of the introduced corpuscles. The possible combinations are as follows:

RECIPIENT'S BLOOD TYPE
(Antibodies in parentheses)

		AB (none)	A (β)	B (α)	O ($\alpha\beta$)
	AB (AB)	o	+	+	+
Donor's blood type. (Antigens in parentheses)	A (A)	o	o	+	+
	B (B)	o	+	o	+
	O (none)	o	o	o	o

A 'o' in the table signifies no reaction while a '+' indicates agglutination of corpuscles. It can be seen from the chart that the blood of an O type person can be used in any transfusion. For this reason type O is spoken of as a *universal donor*. A type AB individual can receive blood of any of the four types. For this reason type AB is spoken of as a *universal recipient*.

SUGGESTED READINGS

Multiple alleles are discussed in all textbooks on genetics. The following are a few of many good ones.

King, R. C. 1962. *Genetics.* Oxford University Press.
Sinnott, E. W., L. C. Dunn, and Th. Dobzhansky. 1958. *Principles of Genetics.* McGraw-Hill.

14

Induced Mutations

The origin of mutants has been a mystery since the early days of genetics. In the initial work of Morgan and his associates, stocks of wild-type Drosophila might be kept for generations, and thousands of individuals examined, before a new mutant was discovered. The occurrence of mutations was a spontaneous event that could neither be predicted nor controlled.

Attempts to induce hereditary changes in the chromosomes were made from the very beginnings of Drosophilia genetics. Various agents were tried, such as exposure of the flies to radium, X rays, and many different chemical agents. In one of Morgan's first papers he reported the appearance of several new mutants in the offspring of flies that had been exposed to radium. Several other investigators reported similar results.

Difficulties in Studying Mutation. None of these early experiments was conclusive because of the difficulty of distinguishing induced from spontaneous mutations. This was the problem. Mutants were appearing in stocks not exposed to unusual radiations or to special chemical treatment. Their appearance could not be correlated with any known 'cause,' so they were termed 'spontaneous.' Spontaneous mutations were of very rare occurrence. In the experiments attempting to produce mutations by physical or chemical means, mutations occurred but they also were very rare. Thus, if we expose flies to radium in an effort to produce mutations, and if a mutant form appears among the offspring or later descendants of the irradiated flies, we could never be sure whether radium was the cause or whether it 'just happened.'

Since new mutant genes appear infrequently and most of them are recessive, the mere detecting of them becomes a problem. Suppose, for example, that one autosomal gene in a sperm nucleus mutates. If this sperm then enters an egg the new individual will have one mutated gene from the father and one unmutated gene at the same locus from the mother. The mutant gene will, consequently, be masked by its dominant allele and the observer will see no evidence that a mutation has occurred. Appropriate crosses could be made to produce an individual homozygous for the new mutation if there was some way of knowing which individuals to cross. Since there is no way of knowing this, the alternative would be to make innumerable crosses in the hope of having at least one fly heterozygous for the new mutant gene.

An appreciation of this problem will be gained if you determine the number of crosses that would have to be made to secure homozygous flies starting with a single adult heterozygous for a new mutant gene. If you then wished to measure the mutation rate per one million flies, what would the total number of necessary crosses be?

Muller's ClB Method. Muller (1927) was the first person to give a practical solution to the problem. He was able to do so because he devised an ingenious experiment that gave an easy and accurate measure of mutation rate. His experiments were designed to test the effects of X rays on the induction of mutations. As a control, it was necessary for him to know the spontaneous mutation rate as well. For his purposes Muller developed the so-called **ClB** strain of Drosophila. A **ClB** ♀ contains the **C** inversion on one of her **X** chromosomes, a recessive lethal gene l and a dominant mutant *bar-eye* **B**, both of the latter two within the inverted section of the chromosome.

An *inversion* is a region of the chromosome that has been reversed. If the normal order of genes is **a b c d e f g**, a chromosome with genes in the order **a b e d c f g** would contain an inversion. An inversion is caused by a double break of the chromosome, in this case between **b** and **c** and between **e** and **f**. Following this there is a rotation of the central section, **c d e**, through 180° and a subsequent fusion with the two ends of the chromosome. Inversions were discovered by Morgan's group, and it was found that they have the important effect of reducing or even preventing crossing over between genes in the inverted section of one chromosome and in the corresponding normal sequence of its homologue. In this particular inversion, the **C** inversion, crossing over is entirely prevented. This means that the l and **B** genes will be inherited as a unit. The **B** gene has the sole purpose, in this experiment, of serving as a ready means of recognizing a fly heterozygous for the **ClB** chromosome, since every fly that has *bar* eyes must have one and only one **ClB** chromosome.

If a female heterozygous for a **ClB** chromosome is crossed with a normal male the results are as follows:

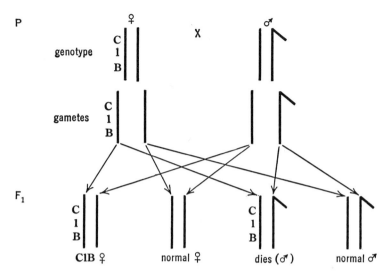

Those males that inherit the **ClB** chromosome from their mothers will have the lethal gene, l. Since there is no normal gene on the **Y** to counteract the effects of this lethal, these males will die. Therefore, the sex ratio will be 2 ♀ : 1 ♂.

It was well known at the time Muller performed his experiment that many separate gene loci can mutate in such a way as to lead to death. (These lethal genes were usually recessive.) Since many genes can do this, the chance of getting a lethal mutation is greater than the chance of observing mutations at specific locus. Thus, if we studied the rate of mutations to the lethal condition on the **X** chromosome we would be measuring the sum of the rates for many different genes. The rate for a specific locus, such as the mutation to *white* eye, would be much smaller.

With **ClB** flies it is possible to measure the rate at which lethal genes appear on the **X** in the following manner. Let us measure the percentage of sperm in which a new **X** chromosomal lethal has occurred. Once again it must be emphasized that this will not be a measure of rate for one locus, but of all the loci on the **X** that can mutate to a lethal condition. The experiment would begin with the crossing of many wild type males with many **ClB** females. In the diagram an * will indicate a new lethal that occurs on the **X** chromosome of the sperm of the male. The diagram will show the manner in which it will be detected.

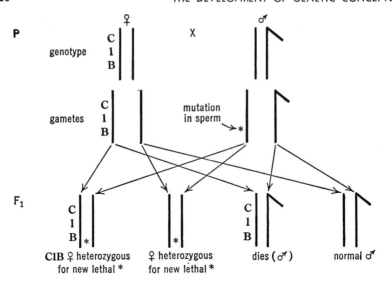

CIB ♀ heterozygous   ♀ heterozygous   dies (♂)   normal ♂
for new lethal *   for new lethal *

If the F_1 CIB ♀ is now mated with a normal ♂, the marked chromosome that we are searching for will pass to the ♂ offspring and be revealed, as shown in the diagram.

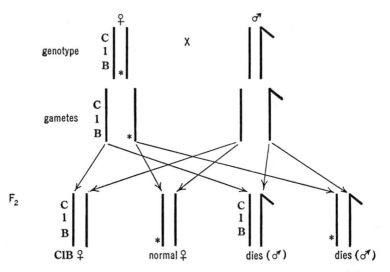

CIB ♀   normal ♀   dies (♂)   dies (♂)

One class of the daughters will be normal in appearance and heterozygous for the new lethal mutant gene. Another class will be CIB females. One of the males will carry the new lethal mutant gene and die as a result. Another class of males will inherit the CIB chromosome and die because of the lethal gene in the C inversion. As a result only females will appear. *Thus, if a lethal gene was present in the sperm of the original male, there will be no sons in the F₂.* This fact can be ascertained by a quick examination of the F_2 flies. This point is of considerable significance, since it makes it feasible to check many crosses in a short period of time.

When Muller began the experiment with normal untreated males, he found that approximately one cross in a thousand gave solely females in the F_2. This means that the chance of a lethal mutation occurring at some locus on the **X** is 1 in 1,000 or 0.1 per cent. This is the natural, or spontaneous mutation rate. If the males are first exposed to about 4,000 r-units of X rays the results are strikingly different. After experimental treatment of this sort approximately 100 crosses in every 1,000 have only females in the F_2. This is a combined mutation rate for all lethal loci of 100 in 1,000 or 10 per cent.

Muller's results were not only of great theoretical importance in showing that mutations could be experimentally produced, but they gave geneticists a practical means of securing new mutants for their work. Later it was found that radiations would induce not only gene mutations but also cause inversions, translocations (the attachment of a piece of one chromosome to another), or deficiencies (elimination of a section of a chromosome).

SUGGESTED READINGS

Muller, H. J. 1927. 'Artificial Transmutation of the Gene.' *Science 66*:84-7.
Muller, H. J. 1928. 'The Production of Mutations by X-rays.' *Proceedings of the National Academy of Sciences 14*:714–26.

15

Salivary Gland Chromosomes

The concepts of genetics were based largely on breeding experiments. That is, the localization and behavior of genes were studied without the gene ever being seen. With the techniques available, the chromosomes of most organisms appeared as uniformly staining structures with no differentiations recognizable as genes. By 1930 geneticists felt that genes were some type of protein molecule. If this were so, it would be impossible to see them even under the most powerful compound microscopes available since molecules are too small to be observed with these instruments. Geneticists became resigned to investigating their invisible genes just as the chemist studies his invisible molecules and the physicist his invisible sub-atomic particles.

The Discovery of Salivary Gland Chromosomes. It was against this background that Painter, in 1934, and somewhat later Bridges, made discoveries of the first importance relating to the finer structure of chromosomes. It was found that the chromosomes in the salivary glands of larvae of Drosophila were of enormous size, being about 100 times longer than those of ordinary body cells. Of even greater interest and importance was the presence of cross bands on the chromosomes. Figure 15–1, from Painter's first paper on the subject, shows the appearance of the salivary chromosomes. It was found that the chromosomes are attached to a chromocenter by their centromere regions. The **X** chromosome has its centromere at one end, and appears as a single structure radiating out from the chromocenter. The second and third chromosomes have their centromeres near the middle, and as a result they have right and left arms extending from the chromocenter. The tiny fourth chromosome appears as a bump protruding from the chromocenter.

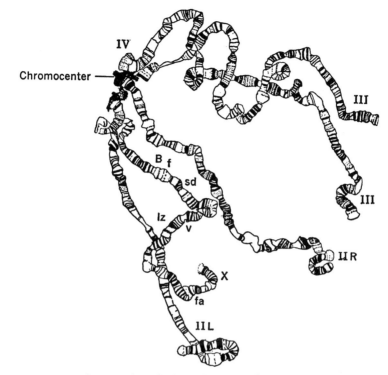

15–1 Painter's first drawing of the salivary gland chromosomes of *Drosophila melanogaster*. The chromosomes radiate out from the chromocenter. The X is attached to the chromocenter by one end so it appears as a single long structure. Both the II and III chromosomes are attached by their middle portions. Consequently both of these chromosomes have two arms extending from the chromocenter. The tiny IV chromosome is attached by its end to the chromocenter. The approximate location of several X chromosome genes (*B*, *f*, *sd*, etc.) is shown. (T. S. Painter, 'A New Method for the Study of Chromosome Aberrations and the Plotting of Chromosome Maps in *Drosophila melanogaster*,' *Genetics* 19:175–88. 1934.)

The salivary glands are diploid in common with all body cells, but instead of the expected eight chromosomes, Painter found only four. This is due to the fact that homologous chromosomes have synapsed. The chromosome labeled **X** is composed of two chromosomes fused together so closely that the line of separation is not shown in the illustration. The line of separation can be seen in good microscopic preparations, however. Note that the pairing of the two chromosomes is so exact that the cross bands extend across both chromosomes as though they were a single structure.

In different sections of the chromosomes the bands were found to differ in size, distinctness, shape, and distance from adjacent bands.

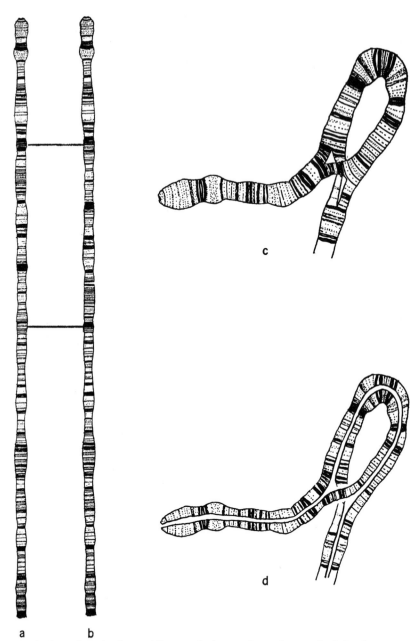

a   b

15–2   Inversions in *Drosophila pseudoobscura.* In *a* the standard band sequence in the third chromosome is shown. The chromosome shown in *b* is identical to the one in *a* except for the region between the two lines, which has the band sequence reversed. This is a cytological demonstration of an inversion. If an individual fly has one of each of these chromosome types, the salivary gland pic-

These regional differences were found to be constant for the same chromosome in different flies. This was of great importance since by means of the bands it was possible for the first time to recognize the various regions of the chromosomes. The genetic map of a chromosome was nothing more than a way of expressing linkage groups and crossover percentages, but now the concepts of chromosome structure based on genetic data could be checked.

Was there any relation between the genetic map and the real chromosome? When geneticists assumed that a group of genes was reversed from the usual order, this was a hypothesis devised to explain the breeding data. Did these genetic inversions mean that there was a chromosomal inversion as well? When a gene changes its linkage group is there an actual translocation of material from one chromosome to another? These questions could be answered through a study of salivary gland chromosomes and many new questions were raised. What is the relation of the bands to genes? If the position of genes in the salivary chromosomes can be established, what is the relation of this to the genetic maps of chromosomes?

Inversions and Translocations. First we shall consider the case of inversions. In the early days of Drosophila genetics, Morgan's group encountered some unusual genetic phenomena. They could be interpreted if it were assumed that the order of some of the gene loci was reversed. This could happen if a chromosome broke in two places, to give three pieces, and the middle piece rotated 180° and then rejoined the two end sections of the chromosome. It was quite an intellectual achievement to make this hypothesis, which at first must have seemed most improbable. Direct evidence for this hypothesis could not be obtained so long as it was impossible to detect regional differences in chromosomes. When salivary chromosomes were discovered, a method of checking became possible.

Figure 15–2 gives an example. In *Drosophila pseudoobscura* the 'standard' arrangement of bands is shown in *a*. In this species many inversions have been discovered. One of these is known as 'arrowhead.' A chromosome with the arrowhead inversion is shown in *b* and it can be seen that the sequence of bands is reversed. Cytology confirmed the genetic hypothesis of chromosomal inversions.

---

ture will be as in *c*. Pairing is achieved by one chromosome forming an inverted U and the other a loop. As a result of these contortions, it is possible for the corresponding bands of the two chromosomes to be situated opposite to one another. In *d* the two chromosomes are separated slightly to show more clearly the manner of pairing. (Modified from Dobzhansky and Sturtevant, 'Inversions in the Chromosomes of *Drosophila pseudoobscura*,' Genetics 23:28–64. 1938.)

An interesting situation arises when an individual has one standard and one arrowhead chromosome. During synapsis these two chromosomes will pair, with corresponding bands being adjacent as shown in *c*. Since a section of one chromosome has a reversed order of regions, considerable gyrations are necessary for matching of the corresponding regions to be achieved. The way it is accomplished is shown in Fig. 15–2*d*.

The occurrence of translocations, which had been predicted on the basis of genetic results, was also verified by an examination of the salivary chromosomes. In those individuals suspected of having translocations, it was found that a piece of chromosome, with its characteristic set of bands, was no longer in its customary place, but instead it was joined to another chromosome.

How did Painter decide which salivary chromosome corresponded to a particular linkage group? It was done largely by means of inversions and translocations. Painter had studied the salivary gland chromosomes of normal flies so carefully that he was able to recognize the various regions. Once he could do this, he studied flies, which were known on genetic grounds to have an inversion in one chromosome— for example, the **X**. Invariably, he found that in one of their salivary chromosomes there was a region of reversed bands. This chromosome was, therefore, the **X**. He had numerous inversions and translocations for study, so it was possible to identify each salivary chromosome with the corresponding linkage group.

The Gene Locus. So much for the gross problems: Now we can ask, Where are the genes? Chromosomal structural aberrations were the basis for solving this problem, small deficiencies being especially useful. The following example, from the work of Demerec and Hoover, will show how it was done (Fig. 15–3). They studied Drosophila with small deficiencies near one end of the **X** chromosome. Most deficiencies, except those that are very small, are lethal when homozygous. In the heterozygous condition they do not cause death, but they do have a characteristic genetic effect, which can be brought out by the following consideration.

Let us assume that a fly is heterozygous for a deficiency including the locus of the gene **A**. That is, there will be no **A** on the chromosome with the deficiency, but in the normal chromosome the locus will be present. The result is that the gene at the **A** locus on the normal chromosome will be the one determining the character of the individual. If the dominant **A** allele is present, it will have its usual effect. If the recessive **a** allele is present it will produce its effect, since there is no gene on the chromosome with the deficiency to influence its action. (These results

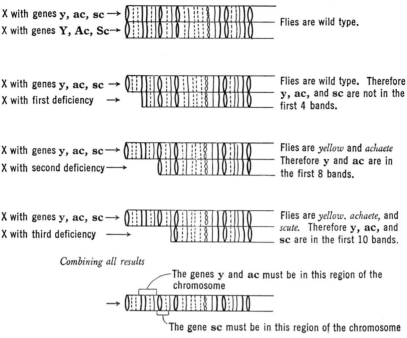

15–3 Diagrammatic representation of the experiment of Demerec and Hoover showing how the positions of genes on the salivary gland chromosomes can be determined.

should not be too surprising. All the **X** chromosome characters of the Drosophila male that we have studied behave in a similar way. The **Y** lacks nearly all loci normally present on the **X** and, therefore, acts as one giant deficiency.)

Demerec and Hoover obtained three stocks, each with a different small deficiency at one end of the **X** chromosome. The locations and extents of the deficiencies were determined by study of the salivary chromosomes. Genetic crosses had established that the genes y (*yellow*), ac (*achaete*), and sc (*scute*) were close to the end of the **X** chromosome. Demerec and Hoover made their crosses in such a way that a fly would receive one normal **X** chromosome carrying the recessive genes y, **ac**, and sc and another **X** carrying a deficiency but no mutant genes. If the deficiency included the locus of any of these genes, then the fly would exhibit the recessive character since there would be no normal allele to counteract its action.

The first deficiency tried was a small one. It removed the first 4 bands of the **X** chromosome. When this deficient chromosome was present with the **X** carrying **y, ac,** and **sc,** the flies were normal. This means that none of these genes are in the part of the **X** having the first 4 bands.

The second deficiency removed the 8 terminal bands. When this chromosome was present with an **X** carrying **y, ac,** and **sc,** the flies were *yellow* and *achaete.* This experiment showed that **y** and **ac** were in that part of the chromosome having bands 1 through 8. The previous cross showed they were not in the region covered by bands 1 through 4. Therefore, the genes **y** and **ac** must be in the region marked by bands 5 through 8. This cross also showed that the locus for **sc** is not in the area covered by the first 8 bands.

The third deficiency removed the 10 terminal bands. When one of these chromosomes was present with an **X** carrying **y, ac,** and **sc,** the flies were *yellow, achaete,* and *scute.* This indicated that **y, ac,** and **sc** were in the area covered by the first 10 bands. In the previous experiment we saw that **sc** was not in the first 8 bands. The present cross shows it is somewhere in the first 10. Combining these results, we can conclude that the **sc** gene is in that region of the chromosome containing bands 9 and 10.

In this manner it was possible to give an approximate location for a number of genes. In a few cases genes were localized to a portion of the chromosome having but a single band. No genes were found in areas without bands. *These observations lead one to the tentative hypothesis that the bands are gene loci.*

If the bands are gene loci, it should be possible to determine the number of genes in Drosophila by counting the number of bands. This attempt was made but one difficulty made an exact determination impossible. Not all of the bands are equally distinct: they vary from those that stain distinctly to those so indistinct as to be at the limit of visibility. Approximately 5,000 bands could be seen and this figure may be taken as the tentative number of genes in Drosophila.

Once genes were located on salivary chromosomes, a comparison could be made with the genetic maps. One example is given in Fig. 15–4, which shows corresponding parts of a small section of the salivary second chromosome and the genetic map. The point of greatest importance is the close resemblance of the two. The genetic map is based on breeding experiments and the arrangement of loci is based on the percentages of crossing over. These data suggest that the genes are in a linear order and in a certain sequence. When it became possible, with the salivary gland techniques, to determine the actual position of genes on the chromosomes, it was found that the order and sequence predicted by genetic means was verified by cytology. Once

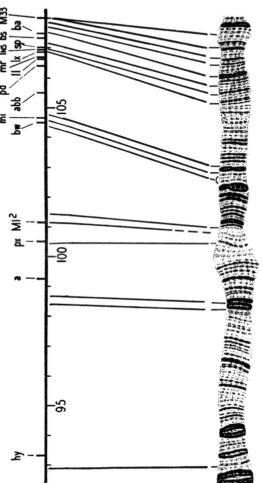

15–4 Corresponding points in the salivary chromosome and linkage map for the tip of the second chromosome of *Drosophila melanogaster*. (Modified from C. B. Bridges, 'Correspondence between linkage maps and salivary chromosome structure, as illustrated in the tip of chromosome 2R of *Drosophila melanogaster*,' *Cytologia*. Fujii Jubilee Volume, pages 745–55, 1937.)

again, these mutually supporting fields had established a hypothesis as 'true' beyond a reasonable doubt.

Salivary gland chromosomes are of great importance in many genetic problems being studied today. One of their more spectacular applications has been in the field of evolution. They will be considered again when this topic is treated.

As an interesting historical footnote we might add this bit of information: The banding of salivary chromosomes in flies had been observed and recorded by cytologists before 1900, but this was not known to geneticists. If the Morgan group had been aware of this, their efforts would have been made much easier. Their prediction of chromosomal aberrations such as inversions, translocations, and deficiencies, largely on genetic data, was a tremendous intellectual achievement. All this time a simple method for demonstrating these cytological phenomena was buried and forgotten in the archives of biological literature.

Cases such as this, which are not infrequent, make all scientists wonder what important facts have been discovered, forgotten, and now await rediscovery and a realization of their worth.

SUGGESTED READINGS

Painter, T. S. 1934. 'A new method for the studying of chromosome aberrations and the plotting of chromosome maps in Drosophila melanogaster.' *Genetics* 19:175–88.
Painter, T. S. 1934. 'Salivary chromosomes and the attack on the gene.' *Journal of Heredity* 25:465–76.

16

The Basic Concepts of Classical Genetics

We have traced the development of the more important concepts of classical genetics. By the late 1930s it was apparent that the mechanisms of inheritance, that is, the laws governing the transmission of genes from parent to offspring, were well understood. When studies were made of inheritance in forms not previously investigated, the results were generally understood in terms already familiar. Many important details were added later, but the main principles were established.

Although there are a small number of main principles, nearly every one of these has its exceptions. In spite of these exceptions, one is left with an impression of the universality of genetic concepts. By and large the same rules seem to hold for plants, animals, and micro-organisms such as bacteria and viruses. An attempt will now be made to list in italics these general concepts and to indicate in roman type some of the exceptions, which for the most part have not been mentioned before.

1. *The basic morphology, physiology, and biochemistry of an organism is determined by its inheritance, acting in a definite environment. That is, by a process of reproduction it has originated from other organisms similar to it.*

2. *Inheritance is the transmission of genes from parents to offspring.* Some of the characteristics of the individual, especially those of the early embryo, are determined solely by the action of the maternal genes during the formation of the ovum in the ovary. Thus the sperm enters an ovum that has some of its characters, such as size, color, and rate at which mitosis will occur, already determined. The paternal genes will not affect these characters. The paternal genes will affect the ova produced when the individual reaches maturity. The effect of the paternal genes on the early embryo, therefore, may be delayed for a generation.

3. *Genes are situated on chromosomes.* There are few exceptions to this. A fraction, possibly very small, of inheritance is dependent upon non-chromosomal structures such as plastids in plants and some virus-like bodies in animals and plants. Even the genes of bacteria and viruses are parts of chromosome-like structures.

4. *Each gene has its particular place, or locus, on a chromosome.* In some instances the position of the locus may be changed as in inversions and translocations, which may shift the position of one or more genes in relation to other genes.

5. *Each chromosome has many genes, and these are arranged in a linear order.* As an exception to this we should remember that some chromosomes, such as the Y of Drosophila, have only a few genes.

6. *The cells of an animal, except those in the process of ova or sperm formation, contain two of each kind of chromosome (diploid condition). That is, all chromosomes are present in homologous pairs, and each locus will be represented twice.* There are several well known exceptions to this statement. a. In some species there are differences in chromosome number between different kinds of individuals. In bees, for example, the females (queens and workers) are diploid and the males (drones) are haploid. b. It is generally believed that an individual will have the same number of chromosomes in all of its cells (except the germ cells) but some exceptions to this are known. In the liver cells of some species of vertebrates, different classes of cells exist. Some have the expected diploid number but others may have twice this number. c. Sex chromosomes, as in the case of **XO** males, offer still another exception to chromosomes existing in homologous pairs.

7. *For each mitotic cycle every gene is duplicated from the chemical substances present in the cell. Cellular reproduction involves a concurrent genic reproduction.*

8. *Genes are capable of existing in several states, each having detectable effects.* The change from one such state to another is known as mutation. The different states of a gene are known as alleles. In spite of this ability to change, genes are very stable. On the average, a gene might be expected to duplicate at least a million times before a mutation occurs.

9. *Genes can be transferred from one chromosome to the homologous chromosome.* This process, which is known as crossing over, is a normal event in meiosis. However, crossing over does not occur in some instances—for example, in the male of Drosophila. Genes can also be transferred to non-homologous chromosomes as in the case of translocations.

10. *Every gamete receives one chromosome of each homologous pair. The distribution of chromosomes to the gametes is a matter of chance.* Thus, each type of chromosome in every pair of homologous chromosomes will be distributed to 50 per cent of the gametes. In the case of males with **XO** sex chromosomes, half of the gametes will receive no sex chromosome at all. This is a complication, though not an exception to the rule as stated.

11. *The distribution to the gametes of the chromosomes of one homologous pair has no effect on the distribution of the chromosomes of the other pairs.* There are, however, a few cases known in which the chromosomes enter the gametes in specific groups.

12. *Fertilization consists of the random union of male and female gametes, each with one chromosome of every homologous pair.* Therefore, *the zygote receives one chromosome of each homologous pair from its father and one from its mother.* Once again, some sex chromosomes, as in **XO** males, introduce a complication.

13. *When the cells of an organism contain two different alleles of the same gene (heterozygous condition), one allele (the dominant) has a greater phenotypic effect than the other (the recessive). In most cases of this sort the heterozygote is nearly identical with individuals homozygous for the dominant alleles.* In a few exceptional cases, the heterozygotes are intermediate in appearance between the homozygous dominant and homozygous recessive types.

14. *Genes produce their effects through the production of chemical substances, which in turn control the biochemical reactions of the cell.* Many biologists believe that each gene controls the production of a specific enzyme, which in turn controls a specific biochemical reaction. This is a fruitful hypothesis, though unproved before 1940.

These simple propositions are thought to be true, with few exceptions. From them, one can deduce the bulk of classical genetics. Classical genetics is indeed an important part of biological science. Since it is concerned with the origin of living things from other living things, it is almost by definition the core of theoretical biology. Pre-1940 genetics was also important for another reason: it had the most logically consistent and complete theoretical structure of any branch of biology. When other branches of biology were struggling to become "scientific," genetics had reached a level of precision characteristic of the physical sciences.

It is doubtful that, however true, this made many geneticists complacent. What was, in fact, their tremendous intellectual achievement? They had reached these conclusions: the similarities and differences among organisms have a genetic basis; genes are parts of chromosomes and the movement of the latter therefore determines the distributions of the former; usually genes are identical generation after generation, but rarely do they mutate.

But what are the functions of genes? When an ovum with an **X** chromosome containing a gene for white eyes is fertilized by a sperm with a **Y** chromosome, what is the functional relation between the presence of this gene and the white eyes of the adult fly? In short, how do the genes exert their effects? There were plenty of hypotheses and some, we now know, were close to the current version of what is "true." There was no sure way, however, to distinguish those that were true from those that were false. The techniques of classical genetics were simply inadequate to deal successfully with the problems of gene action.

But other techniques and approaches were found and, in the years following the early 1940s, progress in genetics has been incredible, to the extent that it represents one of the brighter chapters in man's intellectual history. And it all began through the study of sick mice!

SUGGESTED READINGS

Beadle, G. W. 1945. 'Biochemical genetics.' *Chemical Reviews 37*:15–96.

Darlington, C. D. and K. Mather. 1949. *The Elements of Genetics*. Allen & Unwin.

Haldane, J. B. S. 1954. *The Biochemistry of Genetics*. Allen & Unwin.

King, R. C. 1962. *Genetics*. Oxford University Press.

Muller, H. J. 1947. 'The gene.' *Proceedings of the Royal Society B 134*:1–37.

Sinnott, E. W., L. C. Dunn, and Th. Dobzhansky. 1958. *Principles of Genetics*. McGraw-Hill.

Snyder, L. H. 1951. *The Principles of Heredity*. Heath.

Srb, A. M. and R. D. Owen. 1952. *General Genetics*. W. H. Freeman.

Sturtevant, A. H. and G. W. Beadle. 1939. *An Introduction to Genetics*. Saunders.

17

The Substance of Inheritance

Pneumonia in man is commonly due to the bacterium, *Diplococcus pneumoniae,* known also by an older name—pneumococcus. Before the days of sulfa drugs and antibiotics, it was one of man's most serious diseases. Diplococcus also produces disease in monkeys, rabbits, and mice. Horses, swine, sheep, dogs, cats, guinea pigs, chickens, and pigeons are resistant.

This species shows a large amount of variability, much of which is now known to be genetic. For example, there are many dozen different types, usually designated by roman numerals. In the United States Type I and Type II are the most common. Types III and IV, as well as all the others, are less common. The various types seem to be identical so far as their cellular structure is concerned. Their specificity is due to the chemical composition of the capsule that surrounds the bacterial cells. The capsule is a thick, slimy, polysaccharide (a complex carbohydrate resembling the starch of plants).

The types are identified by their reaction to antibodies. Thus, if Type II cells are injected into a rabbit, the rabbit will form Type II antibodies. These antibodies will react with the polysaccharide capsules of Type II cells but not with those of the other types.

When capsulated cells are grown on culture plates they form colonies that are smooth and shiny in appearance. Mutations that give rise to cells lacking capsules are rather common. Such cells form colonies that are rough in appearance due to the lack of the polysaccharide capsule. The reverse mutation, of cells lacking capsules giving rise to cells that possess them, occurs rarely. Apart from mutations, the characteristic capsules are inherited from cell generation to cell generation. In the same way, cells lacking capsules continue, generation after generation, to give offspring that lack capsules.

There is an important biological property based on the presence or absence of the capsule. When capsules are present, the cells produce disease; when capsules are lacking the cells are harmless. Thus mice inoculated with capsulated bacteria develop a severe bacteremia and nearly always die in a few days. Non-capsulated cells can be injected and the mice remain healthy.

An experiment on these bacteria that was eventually to usher in a new era in genetics was reported by F. Griffith in 1928. Griffith was a Medical Officer working for the British Ministry of Health. His interests in Diplococcus, as judged by his publications, were entirely medical. He gave no suggestion of the tremendous implications that his work was to have for genetics. This is easy to understand, since in 1928 it was not believed that the variations observed in bacteria were comparable to the genetically controlled variations of higher organisms. Furthermore, the medical profession was nearly wholly ignorant of genetics and geneticists were yet to begin a study of inheritance in bacteria.

Griffith began one of his experiments with a culture of Type II bacteria that possessed capsules. Mice died if injected with these cells. As was usual, when these virulent bacteria were grown on culture plates, they produced the smooth colonies characteristic of capsulated cells. After repeated culturing, a few rough colonies appeared. We now know that these were due to a mutation that results in the loss of the ability to synthesize the polysaccharide coat that is the capsule. Thirty mice were injected with these capsuleless bacteria but no bacteremia occurred. The bacteria were harmless.

Griffith observed, as had others before him, that only living capsulated cells would produce bacteremia in mice. If, for example, he killed the capsulated bacteria with heat, they could be injected into mice and no disease would result. It was established, therefore, that the capsular material itself would not produce disease.

Next Griffith made a double injection. His mice received living capsuleless Type II bacteria and heat-killed capsulated Type II as well. On the basis of the data given so far, we might predict that the mice would remain healthy. After all, they were receiving the harmless variant of living bacteria and dead cells of the virulent strain. Yet all of the four mice injected died after five days. Upon examination, their blood was found to be rich with Type II capsulated bacteria. So far as could be determined, they were the same as other strains of capsulated Type II cells.

This was an almost unbelievable result. It appeared that the ability to synthesize a capsule had been transferred from the dead cells to the living cells. Any geneticist of 1928, who might have known of these experiments, would have shuddered. Whatever the nature of the change, it seemed to have nothing to do with the rules that governed inheritance in peas, Drosophila, man, and every other species that had been studied.

How could the results be explained? Possibly the first explanation that comes to mind would be the occurrence of a mutation of a gene lacking the ability to synthesize a capsule to an allele which could. Though possible, this seems unlikely. Recall that the thirty control mice, which were injected with capsuleless cells, did not die. Yet all four mice receiving living capsuleless cells plus heat-killed capsulated cells died.

Another experiment showed that mutation was not the explanation, and also shed an entirely new light on the problem. Once again the experiment consisted of giving the mice a double injection: living capsuleless bacteria plus dead capsulated bacteria. This time, however, *the cells were of different types:* the living capsuleless cells were of Type II, but the dead capsulated cells were of Type I. Eight mice were injected. Two died, one on the third and one on the fifth day. Numerous bacteria were in the blood of these two famous mice and, when cultured and tested, they were found to be Type I capsulated cells. Somehow the Type II capsuleless cells had been converted into Type I capsulated cells. This was not a temporary change. The cells were cultured generation after generation, and they remained Type I. The change was permanent and hence, in the broad sense, genetic. Griffith did not use such terms, but we would say today that one specific genetic type had been converted into another specific genetic type. The change was not a mutation but was apparently due to some influence of the dead cells—strange genetics indeed.

Variations of the experiment gave similar results. Thus, Type II non-capsulated cells were converted to Type III capsulated cells, and Type I non-capsulated cells were converted to Type III capsulated cells. The changes occurred only when living non-capsulated cells plus dead capsulated cells were injected into mice. Griffith was unable to observe the same change when the experiments were carried out in test tubes. The transformation occurred in a living mouse but not *in vitro*.

If we consider the tremendous medical problems caused by *Diplococcus pneumoniae,* it is not surprising that bacteriologists throughout the world were studying its biology. M. H. Dawson, of The Rockefeller Institute for Medical Research in New York, was one of these. In fact, he was doing experiments similar to those of Griffith in England. In 1927 Dawson, together with O. T. Avery, confirmed the even earlier observations that, when non-capsulated cells were injected into mice, capsulated cells would usually be produced. One of their observations was: 'In all cases in which transformation has been effected, reversion has invariably been toward the specific form from which the [capsuleless] form was originally derived.' This was 1927, a year before Griffith had published the results of his experiments using living and dead cells of different Types.

In 1930, Dawson reported that he had repeated and confirmed Griffith's experiments of 1928. He refined the experiments in important ways in order to ensure that the strange observations, though unexplainable, were true. One of his improvements was to begin the strains of bacteria from single cells, rather than use many cells as Griffith had done. A single cell was allowed to reproduce and form a large population. Since this entire population had a single ancestor, and reproduction was by asexual means, it should have a high degree of genetic uniformity. With this precaution, one could rule out the possibility that the change from one type to another was not real but due to the use of mixed cultures. Using the strains obtained from single cells, Dawson did the following:

Non-capsulated cells derived from a Type II strain were injected into mice together with heat-killed capsulated cells of Type I, Type III, or Group IV. In each case the non-capsulated cells were transformed into capsulated cells of the type represented by the heat-killed cells.

In later experiments, Dawson and his associates were able to produce the transformations *in vitro*. This was a most important discovery, making it far easier to find out what it was in the preparation containing the heat-killed cells that led to the transformations. In 1932 and 1933 J. L. Alloway reported that a crude extract of the capsulated cells would cause the transformation.

The evidence was becoming quite convincing that a chemical substance was responsible for the transformations. It is probable that most workers expected the polysaccharide of the capsule to be the active agent. After all, the polysaccharide was responsible for the Type specificity but, wrote Alloway, the polysaccharide 'when added in chemically purified form, has not been found effective in causing transformation of [non-capsulated] [1] organisms derived from [Diplococcus] of one Type into [capsulated] forms of the other Type. When [non-capsulated cells] change into the [capsulated] form they always acquire the property of producing the specific capsular substance. The immunological specificity of the encapsulated cell depends upon the chemical constitution of the particular polysaccharide in the capsule.

[1] My insertions, in brackets, replace older technical terms with contemporary ones.

The synthesis of this specific polysaccharide is a function peculiar to [cells with capsules]. However, since the [non-capsulated] cells under suitable conditions have been found to develop again the capacity of elaborating the specific material, it appears in them this function is potentially present, but that it remains latent until activated by special environmental conditions. The fact that a [non-capsulated] strain derived from one Type of [Diplococcus], under the conditions defined in this paper, may be caused to acquire the specific characters of the [capsulated] forms of a Type other than that from which it was originally derived implies that the activating stimulus is of a specific nature.'

There is nothing in this long quotation, or in any other writing of this period, to suggest that transformation might be a genetic phenomenon. Alloway and others seemed to regard the phenomenon as some sort of a physiological modification—a perfectly reasonable hypothesis on the basis of the data then available.

Dawson's discovery that transformation could occur *in vitro* and Alloway's discovery that a substance causing transformation could be extracted from the bacterial cells, suggested additional experiments. What was the chemical nature of the transforming substance? Alloway knew that it was not the polysaccharide of the capsule surrounding the cells. A likely guess was that it was a protein, for, in the 1930s, it seemed that nearly every important event that occurred in a living system involved or was controlled by proteins. The answer, however, lay elsewhere.

Work on the problem continued slowly at The Rockefeller Institute and, in 1944, a most important announcement was made. Avery, MacLeod, and McCarty reported that they had obtained the transforming substance in a highly purified state and had established its chemical nature beyond a reasonable doubt. They began with huge amounts of Type III cells, using in some experiments the cells from as many as 75 liters of culture medium. The cells passed through a procedure that involved extraction, washing, precipitation, dissolving, and so on. In the end they had no more than 10 to 25 mg. of the active transforming substance. At frequent steps in the procedure they tested the preparation to make sure that it was still active. The purpose of the test was to see if Type II non-capsulated cells could be transformed into Type III capsulated cells. Their final extract, though small in amount, was still highly active. In fact it was far more active, per unit of weight, than the original mass of cells. What was its chemical nature?

The methods used in purifying the extract should have removed all protein and all fat. As a check, the extract was tested for protein and none was found. Numerous other tests were made, including one for the presence of deoxyribonucleic acid (DNA). The extract was found to be exceedingly rich in this substance.

Tests for the closely similar compound, ribonucleic acid (RNA) gave only weakly positive results. In an effort to check on the reliability of the method, some purified DNA from animal cells was tested for RNA. This animal DNA gave the same weak test for RNA as did the purified transforming substance.

These results suggested that the extract contained a large amount of DNA and possibly some RNA. One could not conclude, definitely, however, that the transforming substance was either compound. After all, other substances might be present and one or several of these be the active principle. Nevertheless a good working hypothesis was: the transforming substance is DNA. The next step was to test the hypothesis.

First a comparison was made of the elemental composition of the purified extract and the elemental composition of DNA. The percentage composition for the extract was: 34.88 carbon; 3.82 hydrogen; 14.72 nitrogen; and 8.79 phosphorus. In 1944 the structure of DNA was not completely known, but the theoretical percentages of these elements were thought to be: 34.20 carbon; 3.21 hydrogen; 15.32 nitrogen; and 9.05 phosphorus. The parallel was striking. Furthermore, the ratio of nitrogen to phosphorus, which was theoretically 1.69 in DNA, was found to be 1.67 in the extract. If we assume that the extract consisted largely of the transforming substance, then the results suggested that the transforming substance *could be* DNA and that it *could not be* protein, nor fat, nor carbohydrate.

Further tests substantiated this view. If we assume that the transforming substance is protein, then its activity should be destroyed by the enzymes that digest protein. Two protein-digesting enzymes that occur in pancreatic juice, trypsin and chymotrypsin, were added to an extract containing the transforming substance. There was no loss of activity.

Another experiment indicated that the transforming substance was not RNA. Other workers had discovered an enzyme, ribonuclease, which destroys RNA. When this enzyme was added to the extract, there was no loss of activity. Avery, MacLeod, and McCarty concluded, 'The fact that trypsin, chymotrypsin, and ribonuclease had no effect on the transforming principle is further evidence that the substance is not ribonucleic acid or a protein susceptible to the action of tryptic enzymes.'

Thus they were reasonably sure of some of the substances that the active principle could not be, but how could they prove the hypothesis that it was DNA? Convincing evidence would be obtained if they could use some agent that specifically destroyed DNA and only DNA. If such an agent destroyed the ability of the extract to transform cells, one could conclude that the transforming substance was DNA.

Four years earlier, two investigators had reported that tissue extracts and blood serum contain an enzyme that breaks down the large molecules of DNA. The enzyme, which was obtained in a crude form, is known today as deoxyribonuclease (DNAase). Avery, MacLeod, and McCarty prepared some of this enzyme and added it to their active extract of transforming substance. There was a complete loss of ability to transform cells. This was most convincing. In addition, many other experiments were carried out and the results were all explainable on the basis of the hypothesis that the transforming substance was DNA.

Preliminary observations suggested that the molecular weight of the DNA was very large—about 500,000. The biological activity of the transforming principle was also quite impressive. Transformation could be induced when the transforming substance was present in a concentration of one part in 600 million.

The following quotation shows how the authors explained their observations that an extract of the DNA of Type III capsulated cells could cause Type II non-capsulated cells to start producing capsules that were specific to Type III.

In the present state of knowledge any interpretation of the mechanism involved in transformation must of necessity be purely theoretical. The biochemical events underlying the phenomenon suggest that the transforming principle interacts with the [non-capsulated] [2] cell giving rise to a coordinated series of enzymatic reactions that culminate in the synthesis of the Type III capsular antigen. The experimental findings have clearly demonstrated that the induced alterations are not random changes but are predictable, always corresponding in type specificity to that of the encapsulated cells from which the transforming substance was isolated. Once transformation has occurred, the newly acquired characteristics are thereafter transmitted in series through innumerable transfers in artificial media without any further addition of the transforming agent. Moreover, from the transformed cells themselves, a substance of identical activity can again be recovered in amounts far in excess of that originally added to induce the change. It is evident, therefore, that not only is the capsular material reproduced in successive generations but that the primary factor, which controls the occurrence and specificity of capsular development, is also reduplicated in the daughter cells. The induced changes are not temporary modifications but are permanent alterations which persist provided the cultural conditions are favorable for the maintenance of capsule

[2] The terms in brackets are substitutions for older terms.

formation. The transformed cells can be readily distinguished from the parent [non-capsulated] forms not alone by serological reactions but by the presence of a newly formed and visible capsule which is the immunological unit of type specificity and the accessory structure essential in determining the infective capacity of the microorganism in the animal body.

It is particularly significant in the case of [the bacterial cells] that the experimentally induced alterations are definitely correlated with the development of a new morphological structure and the consequent acquisition of new antigenic and invasive properties. Equally if not more significant is the fact that these changes are predictable, type-specific, and heritable.

Various hypotheses have been advanced in explanation of the nature of the changes induced. In his original description of the phenomenon Griffith suggested that the dead bacteria in the inoculum might furnish some specific protein that serves as a 'pabulum' and enables the [non-capsulated] form to manufacture a capsular carbohydrate.

More recently the phenomenon has been interpreted from a genetic point of view. The inducing substance has been likened to a gene, and the capsular antigen which is produced in response to it has been regarded as a gene product. In discussing the phenomenon of transformation Dobzhansky has stated that "If this transformation is described as a genetic mutation—and it is difficult to avoid so describing it—we are dealing with authentic cases of induction of specific mutations by specific treatments. . . ." . . .

It is, of course, possible that the biological activity of the substance described is not an inherent property of the nucleic acid but is due to minute amounts of some other substance adsorbed to it or so intimately associated with it as to escape detection. If, however, the biologically active substance isolated in highly purified form as the sodium salt of deoxyribonucleic acid actually proves to be the transforming principle, as the available evidence strongly suggests, then nucleic acids of this type must be regarded not merely as structurally important but as functionally active in determining the biochemical activities and specific characteristics of [the bacterial] cells. Assuming that the sodium deoxyribonucleate and the active principle are one and the same substance, then the transformation described represents a change that is chemically induced and specifically directed by a known chemical compound. If the results of the present study on the chemical nature of the transforming principle are confirmed, then nucleic acids must be regarded as possessing biological specificity the chemical basis of which is as yet undetermined.

Dobzhansky's belief that the DNA was inducing mutations was probably shared by most geneticists. After all, it was clearly established that mutations could be produced experimentally. Muller had demonstrated this, in 1927, by using X rays (Chapter 14). Other investigators subsequently discovered that ultraviolet light and some chemical substances would cause mutations.

There is a difference, of fundamental importance, between the effects of X rays and the phenomenon of transformation in bacteria. X rays, while clearly able to cause mutations, do not produce *specific* mutations. One could not, in a defined experimental procedure using X rays, produce only mutations at the white-eye locus in Drosophila. Instead, the X rays would produce many kinds of mutations. Possibly one would be at the white-eye locus, possibly not. The DNA extracted from capsulated bacteria, in contrast, could produce specific changes in non-capsulated cells. Since these specific changes were inheritable, there were grounds for calling them mutations. One could imagine the mechanism to work somewhat as follows: the bacteria contain a gene, which can be either in the capsule-producing or non-capsule-producing state. We can call them c+ and c−, respectively. DNA extracted from c+ cells has the ability to cause the mutation c− → c+.

The implications of this hypothesis, if true, were enormous. Man appeared to be in a position from which he could control inheritance to a degree never before in his power. He might mold his own species and others of importance to him. To be sure, he could do this for only a single gene and in only one species of bacteria, but this limitation need not be too serious. One of the clear implications that could be drawn from genetics was the fundamental identity of hereditary mechanisms in all organisms. The rules for the inheritance of sepia eyes in *Drosophila melanogaster* and of blue eyes in *Homo sapiens* were found to be the same. After Muller had demonstrated the mutagenic effects of X rays in Drosophila, hospitals became more careful about exposing patients and physicians to these radiations.

Of course, there was no theoretical reason to suppose that the mutation process could not be controlled. Mutation, whatever it was, could only be a physical or chemical event—a scientist can conceive of no other possibility. Therefore any mutation, such as c− → c+, would become controllable once the necessary information about the biology of cells was at hand. Prior to 1940 a geneticist might have predicted that the cell biologist should be able to supply him with the necessary information about the year 2000. It was all the more remarkable, therefore, that the feat was accomplished in 1944.

But the true explanation of transformation in bacteria lay elsewhere.

Bacteria have their health problems too. Microorganisms known variously as the bacterial viruses, bacteriophages, or phages can attack bacterial cells and so disrupt the cell's metabolism that death results. In recent years, one bacterium and its many phage parasites have given important new insights into genetic mechanisms.

The bacterium, *Escherichia coli,* is a harmless inhabitant of the large intestine of man. It can be grown readily in the laboratory. Large populations are easy to obtain since cell division occurs about once every 20 minutes. Thus a geneticist, who would have to wait about 75 years for three generations in man, or six weeks in Drosophila, could observe three generations in *E. coli* in one hour. If *E. coli* is infected with a phage, such as one called T_2, the bacterial cell is killed in about 20 seconds. The main steps are as follows. The phage attacks the cell and produces a profound change in the cell's metabolism. Before the phage appeared, the cell was synthesizing its own specific molecules: bacterial proteins, bacterial nucleic acids, and so on. The phage changes all this. In some manner it assumes control of the cell's synthetic machinery and directs it to produce phage molecules instead of *E. coli* molecules. In about 20 minutes, the bacterial cell will have been forced to make about 100 phage particles. It is at this point that the bacterial cell ruptures and liberates the newly formed phage. Each new phage can infect another *E. coli* cell and repeat the cycle.

Phage particles have genetic specificity. The specificity is shown in many ways: the T_2 phages can grow only in living *E. coli* cells; they have a characteristic structure when they are photographed with an electron microscope; chemically they are comparatively simple, being composed of an outer coat of protein and a core of DNA.

The phage particles that are released by the bursting bacterial cell are, in the vast majority of cases, the same as the particle that first entered. Thus the phages exhibit genetic continuity. It follows, therefore, that a biological system consisting of no more than a protein coat and a DNA core contains all the genetic information needed to direct a bacterial cell to make more T_2 phage. These phage particles are only about one-fifth of a micron in length. They are, therefore, far smaller than the familiar carriers of genetic information—the chromosomes. In this system, so simple that it consists of only two parts, it might be possible to determine the molecular basis of inheritance. Does inheritance in the T_2 phage depend on the protein coat, on the DNA core, or on the interaction of both? A partial answer to this question was provided in 1952 by A. D. Hershey and Martha Chase.

Hershey and Chase used radioactive substances to tag separately the protein coat and the DNA core of the phage. This was possible because of a fundamental chemical difference between the protein and DNA. DNA is rich in phosphorus, but it contains no sulfur. On the other hand, the protein of the phage coat contains sulfur, but little or no phosphorus. In 1952 radioactive isotopes of both phosphorus, such as

P^{32}, and sulfur, such as S^{35}, were available. It should be possible, therefore, to obtain phage with its protein containing radioactive sulfur and its DNA containing radioactive phosphorus.

It was well known that the phage reproduced only in the living cells of *E. coli*. Materials in the cells of *coli*, therefore, were sources of the newly synthesized phage. It was necessary, therefore, to introduce the radioactive markers into the T_2 phage by way of the bacterial cell.

The procedure was as follows. One group of bacteria was grown in a medium that contained small amounts of P^{32}. The phosphorus entered the bacterial cells and became part of the cell's molecules. Another group of bacteria was grown in a medium containing S^{35}. Each group of bacteria was allowed to grow for about four hours and was then infected with phage. The phage entered the bacterial cells and reproduced. The new phage was produced, of course, from the materials in the bacterial cells, which contained the radioactive markers. Thus in one group of phage the protein coats became marked with S^{35} and in the other group the DNA was marked with P^{32}. It would then be possible to trace the movements of the radioactively labelled phage protein in one case and the DNA in the other during the infection of bacteria.

At this point we must digress to mention an important observation concerning the mechanics of infection. Other investigators had found that the T_2 phages were elongate structures with a wide cylindrical 'head' and a narrow cylindrical 'tail' (Fig. 17–1). Photos taken with an electron microscope showed the tail of the phage attached to the cell wall of the bacterium. To these observations, Hershey and Chase added others that seemed to indicate that the phage became attached to the bacterium and then injected its DNA into the cell. If this were true, that only the DNA of the phage entered the cell, it would appear that the DNA alone carried genetic information. Having tagged either the protein or the DNA of the phage, Hershey and Chase were in a position to test this hypothesis.

Bacteria were infected with phage that had their protein coats labelled with S^{35}. A few minutes later the bacteria were put in a Waring blendor. The cells are too small to be injured by the whirling blades. However, the solution was agitated so violently that the phage particles were torn loose from the bacterial cell walls. The cells were then separated from the fluid medium. Both fractions were tested for S^{35}. It was found that 80 per cent of the S^{35} was in the fluid and only 20 per cent was associated with the cells. Nevertheless, the cells were infected: in 20 minutes they burst and liberated a new crop of phage.

In a parallel experiment, other bacteria were infected with phage

17-1 The T_2 phage as photographed with the electron microscope at a magnification of 37,000× (photo by J. S. Murphy. *Journal of General Physiology 36:28*).

that had their DNA labelled with P^{32}. In a few minutes the bacteria, plus phage, were put in a Waring blendor, the phage ripped from the bacterial cells, and then the cells and phage separated. Analysis of the cells and fraction containing the phage gave results precisely opposite to those observed with S^{35}. This time it was found that about 70 per cent of the P^{32} was associated with the cells and only 30 per cent with the detached phage particles. Once again, the phage reproduced in the cells, in spite of the drastic treatment.

Thus infection and phage reproduction occurred when most of the DNA entered the cell and most of the protein coat stayed on the outside. The results were somewhat equivocal, for all of the S^{35}, which was the marker for protein, had not remained on the outside. Nevertheless it was not unreasonable to advance this working hypothesis: the phage DNA carries all the genetic information needed for phage replication.

Another way to test this hypothesis is based on the following argument. The hereditary substance is undoubtedly more stable than other substances in the organism. We should expect, therefore, that it would persist intact generation after generation while any non-hereditary materials would not. If the P^{32} is associated with the hereditary material but the S^{35} is not, one would predict different behaviors for them.

Consider first the case of a phage particle, with its DNA marked with P^{32}, infecting a cell. If Hershey and Chase were correct, the DNA alone would enter the cell. Reproduction would begin, and the original DNA would be divided among the daughter phage particles and become diluted, so to speak, but not diminished in total amount. After the infection cycle was completed, and the cell ruptured, the 100 liberated phages should have among them the original DNA of the entering phage. Any hereditary substance should be expected to behave in this way. If the protein were a hereditary material, it should behave in the same way; if not we might expect that only a small portion of the S^{35} that entered the cell would appear in the progeny.

Hershey and Chase put these ideas to experimental test. They found less than one per cent of the S^{35} of the initial phage was recovered in the daughter phages. Other investigators (including J. D. Watson of whom we shall hear more shortly) had just reported that about 50 per cent of the P^{32} that first entered the bacterial cells was recovered in the daughter phage particles.

Hershey and Chase concluded, 'Our experiments show clearly that a physical separation of the phage T_2 into genetic and non-genetic parts is possible. . . . The chemical identification of the genetic part must wait, however, until some of the questions asked above have been answered.' These questions were: '(1) Does any sulfur-free phage material other than DNA enter the cell? (2) If so, is it transferred to the phage progeny? (3) Is the transfer of phosphorus (or hypothetical other substance) to progeny direct—that is, does it remain at all times in a form specifically identifiable as phage substance—or indirect?'

Hershey and Chase were showing necessary and commendable caution in interpreting their remarkable experiments. The implication of their experiments was clear, however: DNA *is* the substance of inheritance in T_2 phage. When one remembers the near universality of genetic phenomena, it is not too difficult to extend the hypothesis to cover all organisms. DNA is the substance of inheritance, the chemical compound of which the genes are composed.

With this hypothesis in mind, we can reinterpret the experiments on transformation in Diplococcus. Avery and his co-workers had shown that DNA extracted from capsulated cells could cause non-capsulated cells to form capsules. The mode of action was thought to be as follows: the DNA stimulates a specific gene mutation in the non-capsulated cells. Other explanations are possible, of course. In the light of the work of Hershey and Chase, an entirely different mechanism might be proposed: The extracted DNA consists, in part, of genes that have the ability to cause the cell to synthesize capsules; these genes enter the non-capsulated cells and become part of the genetic machinery of the invaded cells; in their new environment they initiate the synthesis of capsules. Avery's extract, therefore, could be thought of as containing functional genes. These genes could 'infect' bacterial cells in much the same way as phage infect *E. coli.*

The experiments of Avery, MacLeod, and McCarty, and of Hershey and Chase, were outstanding examples in a large body of data that made it increasingly probable that the substance of inheritance is DNA. In a few viruses, such as the tobacco mosaic virus, the closely similar ribonucleic acid (RNA) substituted for DNA. Apart from these few exceptions, all of which are viruses, the most likely candidate for the hereditary molecule was DNA.

Many important observations were made possible by a staining procedure known as the Feulgen reaction. This procedure, first developed in 1924, proved to be a specific stain for DNA. That is, when properly used, only the DNA of cells is stained. It is possible, therefore, to use this technique to localize the DNA in cells. The nuclei of all animals and plant cells that have been examined are rich in DNA. The cytoplasm is never stained, or is at most, stained very feebly. During mitosis the chromosomes stain deeply while in non-dividing cells the nuclei are nearly uniformly stained. Thus the DNA was localized in precisely that part of the cell where the geneticists had unequivocally located the genes (Chapter 12).

Late in the 1940s, A. W. Pollister, of Columbia University, and others perfected a photometric method for measuring the amount of DNA in a single nucleus. It had been found that the Feulgen reaction could be used as a quantitative method for measuring DNA. That is, the amount of dye bound was proportional to the amount of DNA. Cells are stained and put under a microscope. Exceedingly sensitive photo-cells (working on the same principle as the familiar exposure meters of the photographer) are used to measure the amount of light that passes through the nucleus. The amount of light that passes depends on the amount of stain in the nucleus. If the nucleus contains much DNA, the stain is heavy, and little light will pass. This method allows one to measure accurately relative amounts of DNA in different nuclei. The error of the method is no more than 10 per cent.

This method was used by many investigators to measure the DNA in a wide variety of tissues of animals and plants. The basic findings were these: in any species the diploid nuclei of all the somatic cells appear to have the same amount of DNA; after meiosis, however, the nuclei of sperm and ova have only half as much DNA. An exact parallel is found, therefore, between the amounts of DNA and the number of chromosomes.

The sum of all the work surveyed in this chapter is a tremendous thought. The mysterious gene, that can be mapped but not known, was revealed to be an identifiable organic molecule, which could be localized, extracted, and transferred from organism to organism. Clearly the nature of DNA is an exceedingly important subject.

SUGGESTED READINGS

The original papers referred to most frequently in this chapter are as follows:

Alloway, J. L. 1932. 'The transformation in vitro of R pneumococci into S forms of different specific types by the use of filtered pneumococcus extracts.' *Journal of Experimental Medicine 55*:91–9.

Avery, O. T., C. M. MacLeod, and M. McCarty. 1944. 'Studies on the chemical nature of the substance inducing transformation of pneumococcal types.' *Journal of Experimental Medicine 79*:137–58.

Dawson, M. H. 1930. 'The transformation of pneumococcal types.' *Journal of Experimental Medicine 51*:99–147.

Griffith, Fred. 1928. 'The significance of pneumococcal types.' *Journal of Hygiene 27*:113–59.

Hershey, A. D. and Martha Chase. 1952. 'Independent functions of viral protein and nucleic acid in growth of bacteriophage.' *Journal of General Physiology 36*:39–56.

General discussions of the problems raised in this chapter will be found in:

Bonner, D. M. 1961. *Heredity*. Foundations of Modern Biology Series. Prentice-Hall.

Sager, Ruth and F. J. Ryan. 1961. *Cell Heredity*. John Wiley.

18

DNA—Structure and Function

Few chemistry textbooks written before 1950 had much to say about nucleic acids. Little was known of their chemical nature and nothing of their function. Apparently they were always associated with proteins, forming a class of compounds known as nucleoproteins. The history of these compounds began in 1868 when Friedrich Miescher obtained the first crude preparation by extracting used surgical bandages, which were permeated with pus cells. Later Miescher extracted nucleic acid itself from fish sperm. Sperm might be expected to be a rich source of nuclear substances because the nucleus occupies such a large part of the cell. In fact the ratio of nucleus to cell is higher for sperm than for any other cell. Later it was discovered that the thymus gland is also a rich source of nucleic acid. Many chemical studies were made on calf thymus glands, which were obtained from local slaughter houses.

When nucleic acid of the thymus gland was hydrolyzed, it was found to consist of only a few components: adenine, guanine, cytosine, thymine, deoxyribose (a sugar), and phosphoric acid (Fig. 18–1).

Yeast cells were also extensively studied. Their nucleic acid was found to be much like that of the thymus but it differed in having uracil instead of thymine and ribose instead of deoxyribose. Gradually the belief arose, now known to be incorrect, that animal cells have one type of nucleic acid (with thymine and deoxyribose) and plants have another (with uracil and ribose).

The nucleic acids and nucleoproteins remained the orphans of the chemist for so long largely because they had no obvious importance either inside or outside the cell. Other proteins were clearly of enormous importance: some were the enzymes that controlled the reactions of the living cells; others were the hemoglobins that carried oxygen;

still others were hormones, with their dramatic effects on a variety of life processes. In the first third of the twentieth century nucleoproteins were not extensively studied because there was no urgent reason for so doing. The number of scientific problems that *might* be studied is always far greater than the number that *can* be studied—scientific manpower is always insufficient.

The biochemists of this period concentrated largely on problems associated with the release and utilization of energy within the cell. Here was a problem of clear and obvious importance. It was vigorously investigated and, at one level of analysis, essentially answered. By 1950 the numerous reactions, each with a specific enzyme, in the pathway from glucose to the end products, carbon dioxide and water, were thought to be known. Adenosine triphosphate (ATP) had been identified as a key substance in the storage and transfer of energy within the cell. These biochemical pathways were adorned with many Nobel Prizes.

The experiments described in Chapter 17, however, suggested that DNA was vitally involved in inheritance. In 1944 it was established that DNA was the transforming substance in Diplococcus; evidence obtained in 1952 suggested strongly that the entire genetic information of the T_2 phage is DNA. With leads of this sort, it is not surprising that many biologists turned their attention to DNA. Their working hypothesis was: DNA and the hereditary material are one and the same.

The possibilities for gaining new insights into genetic mechanisms were enormous with a hypothesis so specific. The hypothesis linked two fields, chemistry and genetics, creating the possibility of testing deductions of a chemical nature with genetic data. Conversely, genetic deductions could be tested with data on the chemical nature of DNA.

It is worth a brief digression to emphasize the tremendous utility of hypotheses of this type. They have been characteristic of genetics since the early days, and one might even say that they were largely responsible for the rapid progress in the field. Recall that Sutton's basic hypothesis was that genes are parts of chromosomes (Chapter 6). If this was so, then one should observe a parallel between the behavior of chromosomes in meiosis and fertilization and the behavior of the Mendelian factors in inheritance. Sutton found such a parallel and speeded genetics on its road to becoming a science. In later years geneticists and cytologists constantly checked the discoveries of one field against those of the other. Genetic data first suggested the hypothesis of crossing over. A basis for the event was then found in careful studies of the chromosomes during meiosis (Chapter 10). Bridges advanced the hypothesis of non-disjunction on the basis of genetic data and tested his hypothesis by a study of the chromosomes of his experimental material (Chapter 12). The hypothesis that pieces of chromosomes may become inverted was suggested by genetic data and confirmed by a study of the salivary gland chromosomes (Chapter 15). This type of rigorous checking of the hypotheses of one field by the data of another has not been generally possible in biology.

To return to the main argument: we can test our hypothesis about the DNA molecule with the well-established principles of genetics. Similarly we can anticipate that, as knowledge of the chemistry of DNA becomes available, new insights into genetic mechanisms will be obtained. In order to proceed, we shall accept as true the hypothesis that the gene is DNA. The following deduction follows logically: DNA

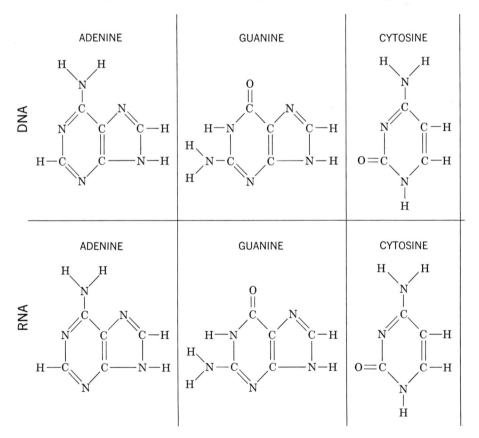

18–1   The hydrolysis products of DNA and RNA.

must have a structure and a composition that will account for the known properties of genes. Let us recall some of these basic properties.

Linkage data localized the gene as part of a chromosome. Experiments on crossing over showed that the genes were in a linear order and in a definable site on the chromosome. Genes were found to be exceedingly stable. Barring mutation, which was a rare phenomenon, the gene maintained its integrity generation after generation. This stability continued even with frequent replication. At each mitotic division every gene became two, and at anaphase one gene went into each daughter cell. A gene might replicate a hundred thousand times or more without making a mistake. Yet from time to time mistakes— or mutations—occurred. Such mutations are essential for the welfare of the species, for they are the raw materials of evolutionary change.

But genes do more than merely maintain themselves. They have

THYMINE	DEOXYRIBOSE	PHOSPHORIC ACID	
			DNA
URACIL	RIBOSE	PHOSPHORIC ACID	
			RNA

specific functions that the geneticist observes in the phenotype of the cell or individual. One gene may play a key role in controlling the synthesis of a polysaccharide capsule on Diplococcus. When an individual possesses a mutated form of the same gene, no capsule is formed. Long before 1950 geneticists had concluded that all the inherited information necessary for life in a specific way—whether as a T₂ phage, Diplococcus, pea, Drosophila, or man—was carried in the individual's genes.

These more important properties of genes can be summarized as follows:

1. Genes must have the ability to make copies of themselves.
2. Genes must have a structure that carries hereditary information.
3. Genes must be able to transfer this information to the rest of the cell.

On the basis of our hypothesis, therefore, the DNA molecule must have a structure that can replicate, carry coded information, and translate this information into the phenotype of the cell, whether the phenotype is a polysaccharide capsule or a specific protein.

THE WATSON-CRICK MODEL

Did DNA have the necessary properties that it must have to be the gene? No one knew in 1950, but an American biologist, J. D. Watson (now of Harvard University) and his English associate, F. H. C. Crick (of Cambridge University) soon addressed themselves to the problem. Could one devise a model of the DNA molecule that would satisfy the requirements imposed by the genetic data?

In two papers, one published in April and the other in May of 1953, they showed how the few facts known about DNA could be used to construct a model of its structure. This model was, in terms of scientific methodology, a hypothesis. Briefly the facts at the disposal of Watson and Crick were these:

1. DNA is composed of six kinds of molecules; adenine, guanine, thymine, cytosine, deoxyribose, and phosphoric acid. These six molecules, however, are combined in DNA in only four ways. Each adenine, guanine, thymine, and cytosine combines with a molecule of deoxyribose and phosphoric acid. The combination adenine-deoxyribose-phosphate is known as deoxyriboadenylic acid. This molecule belongs to a class of compounds known as *nucleotides*. The corresponding nucleotides for guanine, thymine, and cytosine are: deoxyriboguanylic acid, deoxyribothymidylic acid, and deoxyribocytidylic acid.

2. Many of these nucleotides combine to form the huge DNA molecule. (Watson and Crick were attempting to determine the precise nature of their combination.)

3. The available data on the x-ray diffraction patterns of DNA suggested to Watson and Crick that the molecule consists of two long fibers twisted around one another in the form of a helix (like double spiral staircases, one for ascent and one for descent).

4. X-ray data indicated that the diameter of the double helix is about 20 Ångstrom units.

5. Each fiber of the double helix consists of phosphate and deoxyribose units alternating with one another: phosphate-deoxyribose-phosphate-deoxyribose, and so on.

6. The adenine, guanine, cytosine, and thymine units (collectively known as bases) are attached to the phosphate-deoxyribose chain.

7. In different cells of the same species, the relative amounts of adenine, guanine, thymine, and cytosine are the same.

8. In different species the relative amounts of adenine, guanine, thymine, and cytosine vary greatly.

9. In all cells that had been studied, the amount of adenine was found to equal the amount of thymine.

10. In all cells that had been studied, the amount of guanine was found to equal the amount of cytosine. This fact, and the preceding one, was discovered by E. Chargaff of Columbia University.

These were the pieces of a chemical jig-saw puzzle with which Watson and Crick constructed not only a theoretical plan, but a model to determine how the parts of the DNA molecule were put together. Any model would have to account for 'the essential operation required of a genetic material, that of exact self-duplication.' No model for DNA could be seriously considered if it did not account for this clearly established genetic principle.

It is well to emphasize that neither Watson nor Crick had discovered even one of the facts about DNA that have just been listed. These facts had been slowly accumulating over the years and were available to all interested in DNA. It was, however, Watson and Crick alone who were able to see how the data could be unified into a model that would satisfy the genetic and the chemical requirements for the molecular structure of DNA. Theirs was a triumph of the mind, not of the laboratory. Half a century earlier, Sutton had made a similar contribution. Although he did study the chromosomes of grasshoppers, he merely confirmed what others had already established. He saw the relation of the data of Mendel and of the cytologists and combined them to arrive at the hypothesis: genes are parts of chromosomes. Sutton's feat was an exercise of pure intellect, as was that of Watson and Crick. But what *was* their model?

The main part of the long DNA strand was already known to consist of deoxyribose and phosphate linked as shown in Figure 18-2. Other

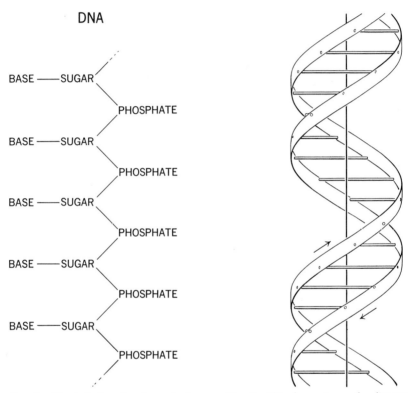

DNA

Fig. 1. Chemical formula of a single chain of deoxyribonucleic acid

Fig. 2. This figure is purely diagrammatic. The two ribbons symbolize the two phosphate-sugar chains, and the horizontal rods the pairs of bases holding the chains together. The vertical line marks the fibre axis.

18–2 Figures 1 and 2 (redrawn) from the original paper by Watson and Crick (*Nature 171:*965).

data suggested that the DNA molecule consists of two of these long strands wound around one another to form a double helix. This much of the model accounts for the sugars and phosphates.

Watson and Crick accepted the earlier belief that the bases are attached to the sugars and more or less stand out at right angles from the long axis of the sugar-phosphate strand (Fig. 18–2). The critical aspect of their model is the positioning of bases on the two entwined strands. Since the relative amounts of the different bases vary from species to species they knew that there could be no single structure for all DNA. They knew also the two striking regularities discovered by Chargaff: the amount of adenine always equals that of thymine and the amount of guanine always equals that of cytosine.

A number of models might provide structurally for these relationships. Watson and Crick constantly kept in mind, however, that their model must provide for self-replication. The model that they finally proposed had the surprising simplicity that characterizes most great hypotheses in biology. Watson and Crick assumed that wherever there is an adenine on one strand there is a thymine opposite it on the other strand and, similarly, that guanine and cytosine are also opposite one another. Thus if we unwind the double helix, the arrangement of the bases will be as shown in Figure 18–3. The adenine and thymine as well as the guanine and cytosine were assumed to be held loosely to one another by hydrogen bonds. These bonds would form if the bases were opposite one another in the positions demanded by the model.

Further evidence for this specific pairing came from data on the relative sizes of the bases and of the diameter of the DNA molecule. Two of the bases, thymine and cytosine, are relatively small. The other two, adenine and guanine, are larger, as can be seen from the diagrams of the molecules in Figure 18–1. Thus the pairing that Watson and Crick assumed was always of one large and one small base. This fit well with the apparently uniform diameter of the DNA, which X-ray data revealed to be 20 Ångstroms. If the pairing were between a cytosine on one strand and a thymine on the other, the diameter would be less than 20 Ångstroms. Similarly if adenine and guanine were to pair, the diameter of the double helix would be more than 20 Ångstroms. The data, therefore, were best explained on the assumption that adenine always pairs with thymine and guanine always pairs with cytosine. (So far as size relations are concerned, adenine could pair with cytosine and guanine could pair with thymine. If this were the case, however, we would expect that the amounts of adenine would always equal those of cytosine and that the amounts of guanine would equal those of thymine. Careful measurements had shown that this was not the case. Instead, adenine always equalled thymine and guanine always equalled cytosine. Once again the data were best explained by assuming a specific pairing of adenine with thymine and guanine with cytosine.)

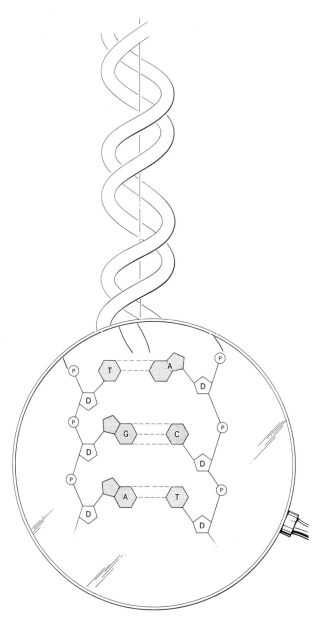

18–3 A highly schematic reconstruction of the double helix formed by the DNA molecule. The lower part of the helix is enlarged to show the bases adenine (A), thymine (T), guanine (G), and cytosine (C) and how these bases are linked with deoxyribose (D), and phosphoric acid (P). Refer to Figure 18–1 for more details of the molecular structures.

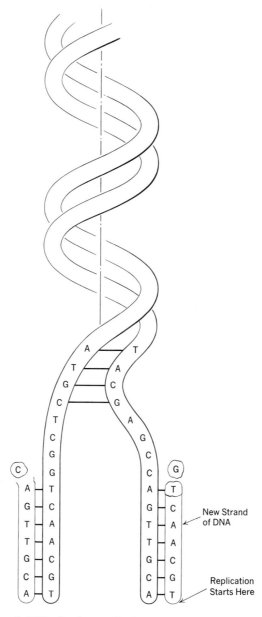

18–4 The replication of DNA. During replication the two strands of DNA are thought to separate from one another. Each strand then serves as a template on which a new strand forms. In this figure replication is beginning at the lower ends of the separated strands. Notice that the end result of replication is two identical double strands.

The Watson-Crick model for DNA, then, consists of two long and closely associated strands wound around one another. The strands are complementary to one another, in the sense that what is present on one strand automatically specifies what is on the other. Thus if the sequence of bases on one strand is adenine-adenine-cytosine-thymine-guanine-thymine, that of the other would have to be thymine-thymine-guanine-adenine-cytosine-adenine. Since the sugar-phosphate parts of the molecule are always the same and only the sequence of the bases can vary, Watson and Crick hypothesized 'it therefore seems likely that the precise sequence of bases is the code which carries the genetical information.'

The gene can make an exact copy of itself; if DNA is the gene it must have the same ability. This was clearly possible with the Watson-Crick model and the argument was developed as follows: 'Previous discussions of self-duplication have usually involved the concept of a template, or mould. Either the template was supposed to copy itself directly or it was to produce a "negative," which in its turn was to act as a template and produce the original "positive" once again. In no case has it been explained in detail how it would do this in terms of atoms and molecules. Now our model for deoxyribonucleic acid is, in effect, a *pair* of templates, each of which is complementary to the other. We imagine that prior to duplication the hydrogen bonds [between the bases opposite to one another in the two strands] are broken, and the two chains unwind and separate. Each chain then acts as a template for the formation on to itself of a new companion chain, so that eventually we shall have *two* pairs of chains, where we only had one before. Moreover, the sequence of the pairs of bases will have been duplicated exactly. . . . Despite [some] uncertainties we feel that our proposed structure for deoxyribonucleic acid may help to solve one of the fundamental biological problems—the molecular basis of the template needed for genetic replication. The hypothesis we are suggesting is that the template is the pattern of bases formed by one chain of the deoxyribonucleic acid and that the gene contains a complementary pair of such templates.' Figure 18–4 shows in diagrammatic form how the replication is thought to occur.

In this way the Watson-Crick model accounts for the important genetic fact that genes can make exact copies of themselves. The model was a hypothesis, which opposed clearly no known chemical or genetic facts. During the next decade an ever-increasing amount of chemical and genetic data suggested that the hypothesis was indeed correct. In 1962 Watson and Crick shared the Noble Prize with Wilkins, the physicist from Cambridge University who had supplied the X-ray data indicating that DNA is a double helix with a uniform diameter of 20 Ångstrom units.

The other primary attribute of genes, their specific function in the cell, was not discussed by Watson and Crick in their first papers. This part of the problem was studied intensively, however, and after a decade a probable hypothesis is at hand.

Genes and Protein Structure. The hypothesis that the chief function of genes is the control of protein synthesis became increasingly probable during the 1940s and early 1950s. A large body of work, much of it on the mold Neurospora, was best interpreted as indicating that genes produce or control the production of enzymes, which in turn control the numerous biochemical events that occur in all cells (all enzymes are proteins). There was no clear understanding of how genes exert their control; do they produce enzymes and other proteins directly or indirectly? Considerable light was shed on this question by the intensive study of a disease of man—sickle cell anemia.

Throughout much of central Africa, sickle cell anemia is common among the natives. The primary effect of the disease is on the hemoglobin of the red blood cells. When these cells are in capillaries where the oxygen concentration is low, they may change from a round to an elongate or even to a sickle shape. These abnormally shaped cells may clog the capillaries and the smallest arteries. Many are destroyed, which causes the anemia. The number of red blood cells may be as few as two million per cubic millimeter, in contrast to the normal number of five million. Infant mortality is high and few individuals with the disease live beyond 40 years.

Genetic analysis has shown that the disease is caused by an autosomal gene, which is symbolized Hb_1^S (the normal allele is Hb_1^A). Homozygous individuals, $Hb_1^S Hb_1^S$, have the severe anemia already described. Heterozygous individuals, $Hb_1^S Hb_1^A$, are nearly normal, however their red blood cells do show abnormal shapes when subjected to very low oxygen concentrations.

Hemoglobin is obviously an important protein, and a great deal was known about its chemistry by the 1940s. Each molecule consists of about 600 amino acids (of 19 different kinds), arranged in two identical halves. Since the most obvious feature of sickle cell anemia is the abnormality of the red blood cells, it is reasonable to suppose that the hemoglobin of these cells might also be abnormal. Linus Pauling and his associates at the California Institute of Technology began an investigation to see if this was indeed so.

Their material was blood from three types of individuals: normal, $Hb_1^A Hb_1^A$; sufferers from sickle cell anemia, $Hb_1^S Hb_1^S$; and the heterozygotes, Hb_1^A Hb_1^S. The blood was fractionated and the hemoglobin obtained in a nearly pure form. In most features the three hemoglobins were identical. When they were compared in an electrophoresis apparatus, however, striking differences were observed.

An electrophoresis apparatus consists basically of a long tube containing a liquid or semi-solid gel. An electric current is passed through the tube. Any charged substances placed in the gel will move, with the rate of movement depending on the size of the particle and its charge. Using this apparatus, it is frequently possible to separate different kinds of molecules in a mixture. Thus, if the three types of hemoglobin differed in their charges, they could be separated, and thereby shown to be different.

This analytical device shows that normal hemoglobin, which we can call hemoglobin A, differs from sickle cell hemoglobin, which we can call hemoglobin S (Fig. 18–5). Furthermore, heterozygous individuals produce both kinds of hemoglobin.

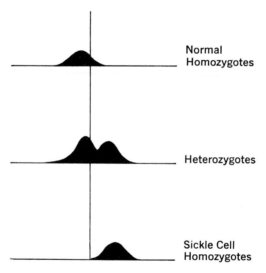

Normal
Homozygotes

Heterozygotes

Sickle Cell
Homozygotes

18–5 Electrophoretic patterns of hemoglobin from normal individuals, from heterozygotes, and from individuals homozygous for the sickle cell gene (modified from Pauling, Itano, Singer, and Wells, 1949).

Pauling's findings are striking evidence that genes can affect the structure of proteins. In the presence of the Hb_1^A gene, the protein hemoglobin A is synthesized in the red blood cells; in the presence of the Hb_1^S allele, hemoglobin S is synthesized.

The next step is to compare the structures of hemoglobin A and hemoglobin S. This is an obvious step, perhaps, but one beset with tremendous difficulties. The problem was to determine the exact position of each amino acid in the total of 600. When Pauling carried out his experiments, the structure of a single protein was not known. It was not until 1954 that F. Sanger, and his co-workers at Cambridge University, finally succeeded in determining the complete amino acid sequence for a protein. After ten years of intensive work they knew the position of each of the 51 amino acids in the insulin molecule.

If it took ten years to determine the structure of insulin with its 51 amino acids, how long might it take to determine the structure for hemoglobin with its 600 amino acids? The task was begun by Vernon Ingram, then also at Cambridge University and now at the Massachusetts Institute of Technology, and his associates. They were able to learn the answer by an analytical short cut—they did not have to determine the complete structure of hemoglobin. This, however, they have since accomplished.

It was nearly impossible to study the huge hemoglobin molecule intact. A more practicable method of investigation was to break down the large molecule into smaller molecules and then to study the smaller molecules. If the structure of each of the smaller molecules could be determined, and if it could then be determined how these smaller molecules are combined, one would know the structure of hemoglobin. The usual tools of the analytical chemist were not of much help. Typically, he hydrolyzes the proteins with acid and gradually breaks down the large molecules until only amino acids remain. The intermediate breakdown products are not uniform, however. A confusing mixture of large and small molecules is formed and this is not suitable for the careful analysis that Ingram was attempting. He required a precise method for breaking down the hemoglobin molecule into smaller molecules. The method he chose was to treat the protein with the enzyme trypsin. Trypsin is highly specific in action, hydrolyzing the protein only in those parts of the molecule where the amino acids arginine and lysine occur. Since these amino acids are always in the same places in the hemoglobin molecule, the hemoglobin will always be broken down in the same way. When the hemoglobin molecule is broken at the site of each arginine and lysine, the result is 28 kinds of smaller molecules, averaging about ten amino acids each. In this manner, Ingram hydrolyzed both hemoglobin A from normal individuals and hemoglobin S from sufferers of sickle cell anemia.

Next he separated the 28 smaller molecules by a process combining electrophoresis and paper chromatography. He put a drop of the hydrolyzed hemoglobin mixture on a piece of paper and an electric current passed through the paper. This method is essentially the one used by Pauling to separate the entire molecules of hemoglobin A and hemoglobin S. In this case, Ingram was attempting to separate the 28 hydrolysis products on the basis of their electrical charges. He achieved considerable separation, but not enough for analytical purposes. Then he tried another method. The edge of the same piece of paper was put in a liquid that would dissolve the 28 hydrolysis products. As the liquid moved through the paper it carried the 28 kinds of molecules with it. The different kinds were carried at different rates. At the end of the experiment, the 28 types of molecules occupied different positions on the sheet of paper—each in a specific place (Fig. 18–6). This type of separation, using both electrophoresis and paper chromatography, gave consistent results. That is, in repeated experiments with hemoglobin A, the 28 spots always occupied the same positions relative to one another.

When hemoglobin S was analyzed in a similar way, again there were 28 spots. But when these spots were compared with those from hemoglobin A, there was an important difference. Twenty-seven of the spots occupied the same relative positions on two sheets of paper, one with hemoglobin A and the other with hemoglobin S. The twenty-eighth pair, however, occupied slightly different positions (Fig. 18–6). The

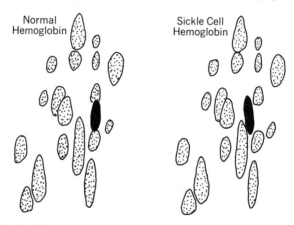

18–6 The hydrolysis products of normal hemoglobin and sickle cell hemoglobin. The products have been separated by paper chromatography and electrophoresis (not all are shown here). In most cases the hydrolysis products of normal hemoglobin and sickle cell hemoglobin occupy equivalent positions. One spot in each, however, occupies a slightly different position. The only difference in the two kinds of hemoglobin is found in this pair of spots (modified from Ingram, 1958).

implication is that hemoglobin A and hemoglobin S are identical in all but one of the units that result from hydrolysis with trypsin. It was, therefore, important to study the one instance of difference. If the structure of this variant could be determined, one would know the difference between hemoglobin A and S without knowing the entire structure of either.

The variant spot for both hemoglobin A and hemoglobin S was analyzed and found to consist of nine amino acids. After much effort, the sequence of these nine amino acids was determined. For hemoglobin A, Ingram found it to be: histidine-valine-leucine-leucine-threonine-proline-glutamic acid-glutamic acid-lysine. For hemoglobin S the order is: histidine-valine-leucine-leucine-threonine-proline-valine-glutamic acid-lysine. These two sequences are identical except for one animo acid. The seventh amino acid in the sequence for hemoglobin A is glutamic acid; for hemoglobin S it is valine. Thus the gene $Hb_1{}^A$ is associated with the production of a molecule that has glutamic acid at one specific site but, in the presence of the allele $Hb_1{}^S$, valine is substituted. When both alleles are present, as in heterozygous individuals, both types of hemoglobin are produced.

This may seem like a trifling difference for a molecule consisting of 600 amino acids (since each molecule of hemoglobin consists of two identical halves, the valine substitution has occurred in two places in the entire molecule). Nevertheless, it alters the hemoglobin molecule to such an extent that it is a lethal or semi-lethal condition in homozygous individuals.

The link between the gene, or DNA, and protein was now even more secure, but its nature was largely unknown. After all, the genes are in the nucleus and most of the cell's proteins are in the cytoplasm. How, then, could one account for the origin of the cytoplasmic proteins? One could hypothesize that they are produced in the nucleus and then pass into the cytoplasm. Alternatively, one could imagine that some influence, or information, which passes from the nucleus to the cytoplasm, directs the synthesis of cytoplasmic proteins. The answer to these questions came from studies of the other nucleic acid, RNA.

Ribonucleic Acid. During the 1930s methods were perfected for detecting both DNA and RNA *in situ*. Some of the methods depended on the different staining reactions of DNA and RNA. The Feulgen method, already mentioned, was specific for DNA. Another cytological method involved the use of two dyes, methyl green and pyronin. Methyl green was found selectively to stain DNA and pyronin to stain RNA. Spectrophotometric methods were also developed to detect the nucleic acids. The bases in these compounds (and in a few others in the cell, such as ATP) intensely absorb ultraviolet light having a wave length of 260 mμ. There are no other compounds in the cell having this specific absorption. Since the bases are situated largely in

DNA or RNA, a peak absorption at 260 mμ indicates the occurrence of nucleic acids. This method will not distinguish between DNA and RNA; it measures total DNA plus RNA plus any other substances, such as ATP, in the cell that contain the bases. If one knows the total, however, and then measures the DNA by the Feulgen method, the difference will be largely the RNA.

These various methods gave consistent data. The DNA is always localized in the nucleus. Most of the RNA is in the cytoplasm but some is in the nucleus. For example, the nucleolus appears to consist entirely of ribonucleoprotein and small amounts of RNA were often detected in association with the chromosomes.

Once it was known that DNA is of great importance in the cell, it is reasonable to suppose that the closely similar RNA is also of importance. The differences between the two nucleic acids were small: RNA had uracil instead of thymine as one of its four bases and its sugar was ribose instead of deoxyribose (Fig. 18–1).

One of the first hypotheses of the role of RNA in the cell was based on the observation that cells that synthesize large amounts of protein are rich in RNA. In the liver and pancreas, where there is much protein synthesis, there is usually from two to eight times as much RNA as DNA. In the kidney, brain, spleen, and thymus, which synthesize much less protein, there are usually equal amounts, or there may even be more DNA than RNA. Possibly this correlation has a causal significance.

This was the view of Jean Brachet, of the Free University of Brussels. In the mid 1940s he observed repeatedly that cells synthesizing large amounts of protein stain heavily for RNA. The stain seemed to be taken up by tiny granules but these were too small to be studied with the compound microscope. Later, when the electron microscope methods were perfected for studying cells, G. Palade, of The Rockefeller Institute, observed these granules. They were called ribosomes because they were found to be rich in RNA. Within the cell they were situated on the walls of a network of tubes known as the endoplasmic reticulum (Fig. 18–7). Palade agreed with Brachet that ribosomes had something to do with protein synthesis. Further, since ribosomes are rich in RNA, possibly it is the RNA that is concerned with protein synthesis.

If this hypothesis was to be tested, clearly very special methods had to be developed. After all the ribosomes were so small that they could not be seen with the highest powers of a compound microscope. Further progress depended heavily on the development of cell free systems.

Ribosomal RNA and Transfer RNA. In the century following the formulation of the cell theory (Chapter 2), the cell came to be regarded

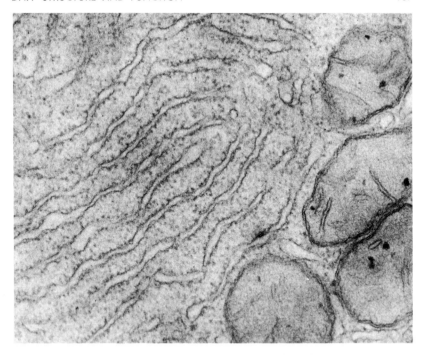

18-7 A portion of a liver cell of a white rat. Photographed with an electron micro-scope at a magnification of 50,000 diameters. The large irregular spheres are mito-chondria. The tube-like structures are portions of the endoplasmic reticulum. The dark granules situated on the walls of the endoplasmic reticulum are the ribosomes, which are the sites of protein synthesis (Photograph by L. D. Peachey).

not only as a unit of structure but also as a seemingly indivisible unit of function. Few of the complex events that occurred in cells could be duplicated apart from them. It was known that enzymes could act *in vitro* and that some small organic compounds could be synthesized from inorganic molecules. But a biochemist, who could easily synthe-size fats, carbohydrates, and proteins in his body, was powerless to accomplish this feat in his laboratory. This inability to reproduce *in vitro* common events occurring *in vivo* did not suggest the need for some vitalistic principle. It was realized that the biochemical events that occur within cells frequently require dozens of different kinds of enzymes, specific sources of energy, and numerous and varied raw ma-terials. The oxidation of glucose, for example, was found to require dozens of enzymes and many complex molecules such as the riboflavins, cytochromes, and so on. There was no reason to believe that even these intricate chains of reactions could not be eventually carried out apart from living cells. It was merely a matter of waiting for the slow accumulation of knowledge and for the perfecting of the necessary methods to reproduce the desired conditions.

Beginning in the late 1940s there was increasing success in isolating cell fractions that were functional. The general method is to grind up, or homogenize, tissues or masses of cells and then to fractionate the homogenate. Usually fractionation is accomplished by centrifuging the homogenate. When this is done, the heavier particles, such as nuclei and unbroken cells, are thrown to the bottom of the centrifuge tube. Smaller and less dense particles form layers above the nuclear layer. For example, a distinct layer immediately above the nuclei contains nearly all of the cell's mitochondria. Above this is a layer consisting almost solely of fragments of the endoplasmic reticulum with the attached ribosomes. The uppermost layer, or supernatant, is a liquid free of all but the smallest particles. The layers are far from pure but, with this analytical procedure, it is possible to obtain relatively pure preparations of mitochondria or ribosomes (methods were developed for removing the ribosomes from the fragments of the endoplasmic reticulum).

Some of these fractions are able to function in limited ways and for limited periods of time. Thus the fraction containing the mitochondria is capable of carrying out oxidative reactions and of forming adenosine triphosphate (ATP), which is the immediate source of energy for all of the cell's reactions.

Possibly the most complex, and certainly one of the most basic, synthesis that occurs in cells—protein synthesis—could be duplicated in cell-free fractions. In this case, the layers containing the ribosomes and the supernatant were involved. The methods were developed largely by Paul C. Zamecnik and his associates at Harvard University and the Massachusetts General Hospital. Their work, together with that of many other investigators, has given a reasonable account of the main steps in protein synthesis, which I shall now summarize.

The cell makes its proteins from 20 kinds of amino acids, which occur in small amounts in the fluid portion of the cell. The first main step is the 'activation,' or combining, of each amino acid with a molecule of ATP. This union is controlled by an enzyme and, apparently, each kind of amino acid requires a different enzyme. Thus there are 20 activating enzymes.

The next step is associated with RNA. This is not the RNA found in ribosomes but relatively small molecules of this nucleic acid, which are dissolved in the fluid portion of the cell. It is called either soluble RNA, because it is dissolved, or *transfer RNA,* because of its function. Each molecule of activated amino acid combines with a molecule of transfer RNA. The data currently available strongly suggest that there is at least one kind of transfer RNA molecule for each of the 20 kinds of activated amino acids. Twenty different enzymes control the formation of the activated amino acid-transfer RNA molecules.

These transfer RNAs with their attached amino acids now move to the surface of the ribosomes. The surfaces are not uniform but possess great topographical specificity, apparently attributable to the RNA of the ribosome. The activated amino acid-transfer RNA molecules line up on the surface of the ribosome, although just how they accomplish this is not yet completely understood. Presumably each specific activated amino acid-transfer RNA molecule can be attached at only a specific site on the ribosomal RNA. Thus a lysine-transfer RNA molecule will attach at one place and a glycine-transfer RNA molecule at another.

Up to this point, each amino acid has been attached to its specific transfer RNA. Now, at the surface of the ribosome, the amino acids break their bonds with the transfer RNAs and combine with one another. In this way proteins are formed. Thus the exact sequence of amino acids in the protein chain is determined by the surface configuration of the ribosome.

These new data can now be added to our general explanation. The fact that the DNA of the gene can determine protein structure can be modified as follows: the DNA determines the structure of the ribosome, which in turn determines the structure of the protein. But the DNA is still in the nucleus and the ribosomes we are discussing are in the cytoplasm (there are other ribosomes in the nucleus). How does the information on how to make a protein pass from the DNA to the surface of the ribosome?

Messenger RNA. Considerations such as these led Crick, in 1958, and François Jacob and Jacques Monod (of the Pasteur Institute of Paris), in 1961, to the hypothesis that a substance must carry the message from DNA in the chromosomes to the ribosomes in the cytoplasm. Theoretical considerations, as well as some data, suggested that the hypothetical substance might be a specific kind of RNA. Therefore the name *messenger RNA* seemed appropriate.

The DNA of the gene, therefore, can be thought of as doing two things. First it can make copies of itself in the manner suggested by Watson and Crick (Fig. 18–4). Thus, if one DNA strand has the base sequence adenine-thymine-adenine-cytosine-guanine-thymine, it can serve as the basis for making the complementary strand consisting of thymine-adenine-thymine-guanine-cytosine-adenine. Second, the same strand of DNA can also serve as a basis for making the messenger RNA. We must remember, in this connection, that RNA has uracil instead of thymine. Thus the base sequence for a messenger RNA made from the DNA of our example will be uracil-adenine-uracil-guanine-cytosine-adenine.

The messenger RNA can be thought of as being synthesized on the chromosomes and then moving into the cytoplasm where it becomes attached to the ribosomes. There were some data suggesting that messenger RNA must have a very short life. That is, whatever information it imparted to the ribosome remains there only briefly. Thus a given ribosome will have the surface structure for making one kind of protein now and, a few seconds later, it could be making either no protein or some other kind of protein. One might think of the surface of the ribosome as 'wearing out' after producing a few molecules of a particular kind of protein. It could be restored by becoming associated with a new molecule of messenger RNA.

The hypothesis of Jacob and Monod, therefore, was this: genetic information is carried from genes to ribosomes by messenger RNA, which has a transitory effect. Based on this hypothesis, they suggested two deductions:

1. Molecules that account for the assumed properties of messenger RNA must be present in cells that are making proteins. The molecules must be polynucleotides, which are produced by the chromosomes, and then become attached to the ribosomes. Furthermore, the base composition of the messenger RNA must correspond to that expected from the base composition of the DNA that makes it.

2. The same ribosomal particle should be able to synthesize different proteins at different times, depending on the instructions it receives from the messenger RNA.

Both of these deductions have been tested and found to be true. Two experiments will be chosen from the many that establish the existence of messenger RNA.

The first of these experiments, performed by M. Zalokar in 1960, is not only pertinent to the argument but also shows the sorts of questions that can be answered with the use of radioisotopes (recall also their value in the experiments of Hershey and Chase, page 144). If one wishes to observe the formation of a substance in the nucleus and its movement into the cytoplasm, special techniques must be used. Presumably these events are rapid, occurring in seconds or minutes, and the amounts of materials involved are too small to be detected by the usual methods of the analytical chemist. In many instances, technical problems of this sort can be surmounted by the use of radioactive isotopes. If one wishes to trace the movements of RNA, it is necessary to mark the RNA in some specific way, so that it can be distinguished from all other substances in the cell.

Zalokar used a substance specific to RNA, namely, uridine (uracil plus ribose) which contains tritium (a radioactive isotope of hydrogen —H^3). The mold Neurospora was his biological material. The mold was given the radioactive isotope for short periods of time, one to four minutes. Then he examined the cells at frequent intervals. During the first few minutes the radioactivity was restricted to the nucleus, indicating that the uracil was located in this part of the cell. After eight minutes the label began to appear in association with the ribosomes. These observations were interpreted as follows: the uracil enters the cell and, in the nucleus, is incorporated into RNA; later this RNA moves into the cytoplasm and joins with the ribosomes. This movement is exactly what one would expect of the hypothetical messenger RNA. Other data obtained with the radioactive uracil showed that at least 99 per cent of the cell's RNA is made in the nucleus and then migrates into the cytoplasm.

The second experiment was reported by E. Volkin and L. Astrachan in 1957. Recall that part of the messenger RNA hypothesis demands that the base composition of the messenger RNA must conform to the base composition of the DNA by which it is formed. Recall also the results when the bacterium *Escherichia coli* is infected with the T_2 phage (page 144): almost immediately the bacterial cells stop making their own specific molecules and begin to make phage DNA and phage proteins. If the messenger RNA hypothesis is correct, the events would be as follows: messenger RNA would be made on the phage DNA instead of on the bacterial DNA; this new and different messenger RNA would move to the ribosomes where it would give the instructions for making phage proteins. If the bacterial DNA and the phage DNA differ in their base compositions, there should be a difference between the RNA produced by an uninfected bacterial cell and that produced by a cell after it has been infected by T_2 phage. This expectation was borne out. After the phage had entered the cell, the synthesized RNA reflected the base composition of the phage, not of the bacterium.

The second deduction of Jacob and Monod, that the ribosome acts as the messenger RNA instructs, was also shown to be true. In 1961 S. Brenner (of Cambridge University), F. Jacob (of the Pasteur Institute), and M. Meselson (then at the California Institute of Technology) worked together at the California Institute of Technology along the general lines of those of Volkin and Astrachan. They also used *E. coli* and T_2 phage. Bacterial cells were given various isotopes and the experiments were designed so that the ribosomes and RNA produced before and after the phage entered the cells could be distinguished. They were able to show that '(1) After phage infection no new ribosomes can be detected. (2) A new RNA with a relatively rapid turnover is synthesized after phage infection. This RNA, which has a base composition corresponding to that of the phage DNA, is added to preexisting ribosomes. . . . (3) Most, and perhaps all, protein synthesized in the infected cell occurs in pre-existing ribosomes.' Thus, 'Ribosomes are non-specialized structures which synthesize, at a given time, the protein dictated by the messenger they happen to contain.'

The Message. The sender of the message (DNA), the messenger (messenger RNA), the receiver of the message (the ribosome), and the consequence of the message (a specific protein) have all been described —but what is the message? The answer to this, one of the most important of all scientific questions, is at hand.

Part of the answer comes from arm chair speculation and part from some extraordinarily sophisticated experimentation. The speculation, which we shall consider first, has consisted, more or less, of playing the 'numbers game.'

Proteins may be in the form of long chains or be coiled, cross connected, or folded in a variety of specific ways. The basis of the specificity lies in the sequence of amino acids that comprise the protein. From the 20 common amino acids the vast majority of all proteins, whether of virus, bacterium, plant, or animal, are made. The data available in 1960 suggested that, beyond a reasonable doubt, the sequence of amino acids is determined by the genes. The genes, in turn, consist of a sequence of bases of the long DNA molecule, which parallels the sequence of amino acids in the protein.

If DNA were composed of 20 different bases, one would suspect that each base would correspond to an amino acid. Thus a thymine in a particular location in DNA might specify that leucine should occupy a specific spot in a protein molecule. Such a scheme cannot work, of course, because there are only four bases in DNA: thymine, adenine, guanine, and cytosine.

It is, however, conceivable that two bases could specify a particular amino acid. Thus thymine-guanine might be thought to be the code for leucine or for some other amino acid. This is also impossible. There can be only 4^2, or 16, combinations of pairs of the four bases—and there are 20 amino acids.

With three bases, however, there are 4^3, or 64, possible combinations. Thus a code composed of triplets of bases would be the minimum number required. The total of 64 is more than three times the number required. There is the theoretical possibility that the code for some amino acids might consist of two bases and that for others of three bases. Possibly the scientists' love of symmetry and order led most of them to adhere to the hypothesis that groups of three bases in the DNA molecule must somehow contain the information for lining up specific amino acids in protein chains.

When the messenger RNA hypothesis became well established, it was possible to think in terms of more precise models. Let us assume, for example, that the sequence adenine-guanine-cytosine is the DNA code triplet for serine. The messenger RNA formed on this part of the DNA molecule contains then uracil-cytosine-guanine (UCG). (The messenger RNA molecule is very large and we are now discussing one small portion of it—a single triplet.) This messenger RNA molecule, with its UCG triplet, becomes attached to a ribosome. Somehow the serine-specific transfer RNA, with its attached serine, reaches that portion of the ribosome containing the UCG triplet of the messenger RNA. The serine becomes detached from the transfer RNA and then attaches to the amino acids adjacent to it on the messenger RNA. These adjacent amino acids would have been brought to the surface of the ribosome in a series of specific events similar to ones described for serine. Eventually all the amino acids adjacent to each other on the messenger RNA join together as a complicated protein, which then separates from the surface of the ribosome. In this manner the code of DNA becomes reflected in the specific amino-acid sequence of the protein.

Hypotheses suggested by this model cannot be tested directly by the methods available today. As is often the case, indirect methods can give the answer.

It had been discovered by M. Grunberg-Manago and S. Ochoa (of New York University Schol of Medicine) that RNA could be made synthetically from mixtures of the four ribonucleotides. The nucleotides of RNA consist of a base plus ribose and phosphoric acid. They are adenylic acid, guanylic acid, cytidylic acid, and uridylic acid. The enzyme, polynucleotide phosphorylase, is necessary to combine the nucleotides to form RNA. Any combination of nucleotides, or even one kind alone, can be used. Thus the enzyme, plus uridylic acid, will form a synthetic RNA composed solely of a long chain of these nucleotides. This particular synthetic RNA is called poly U.

Methods had been perfected, as mentioned earlier, for obtaining the synthesis of proteins in cell fractions. The basic ingredients were the ribosomes and the supernatant. These methods were further refined by M. W. Nirenberg and J. H. Matthaei, of the National Insti-

TABLE 18-1a

The codes for the 20 amino acids have been tentatively determined in experiments with synthetic RNA, as described in the text. Using alanine as an example, the team at the National Institutes of Health made RNA from a mixture of two parts of cytidylic acid and one part of guanylic acid. It is assumed that the RNA so formed consists of twice as much C as G. This RNA promotes the incorporation of alanine into proteins. Other experiments, as well as theoretical considerations, suggest that the code is a triplet. If this is so, the code for alanine consists of two C and one G. The exact sequence is not known; the possibilities are CCG, CGC, or GCC. The precise sequence has been established only for tyrosine—AUU. The codes for alanine, arginine, proline, and threonine all seem to have CCG. This need not mean that the same triplet codes all. There are three possibilities for sequences consisting of CCG, as we have just seen, and proline and threonine can be coded by other triplets as well. (Data modified from *Science 139* page 775. February 1963).

AMINO ACID	TENTATIVE CODE OF MESSENGER RNA						
	AS DETERMINED AT THE NATIONAL INSTITUTES OF HEALTH				AS DETERMINED AT NEW YORK UNIVERSITY		
Alanine	CCG				CCG	CGU	ACG
Arginine	CCG				CCG	CGU	AAG
Asparagine	AAC				AAC	AAU	ACU
Aspartic acid	AAC				AGU	ACG	
Cysteine	GUU	GGU			GUU		
Glutamic acid	AAC	AAG	AGU		AAG	AGU	
Glutamine	AAC				AAC	AGG	
Glycine	GGU				GGU	AGG	CGG
Histidine	ACC				ACC	ACU	
Isoleucine	AUU				AUU	AAU	
Leucine	GUU	AUU	CUU		GUU	AUU	CUU
Lysine	AAA	AAC	AAG	AAU	AAA	AAU	
Methionine	AGU				AGU		
Phenylalanine	UUU				UUU	UUC	
Proline	CCC	CCU	ACC	CCG	CCC	CCU	ACC
Serine	CUU	CCU	CGU		CUU	CCU	ACG
Threonine	AAC	ACC			AAC	ACU	CCG
Tryptophane	GGU				GGU		
Tyrosine	*AUU*				*AUU*		
Valine	GUU				GUU		

TABLE 18–1b

There are 64 possible triplets that can be made from the four bases. However, if one knows only the number of bases involved but not their exact sequence only 20 groups of triplets can be distinguished. These are listed above with the amino acids they are believed to code. Of the possibilities, only GGG seems not to carry information. (These data are based on Table 18–1a.)

INDISTINGUISHABLE GROUPS OF RNA TRIPLETS						AMINO ACIDS CODED
AAA						lysine
AAC	ACA	CAA				asparagine, aspartic acid, glutamic acid, glutamine, lysine, threonine
ACC	CCA	CAC				histidine, proline, threonine
AAG	AGA	GAA				arginine, glutamic acid, lysine
AGG	GAG	GGA				glutamine, glycine
AAU	AUA	UAA				asparagine, isoleucine, lysine
AUU	UAU	UUA				isoleucine, leucine, tyrosine
CCC						proline
CCG	CGC	GCC				alanine, arginine, proline, threonine
CGG	GCG	GGC				glycine
CCU	CUC	UCC				proline, serine
CUU	UCU	UUC				leucine, phenylalanine, serine
GGG						?
GGU	GUG	UGG				cysteine, glycine, tryptophane
GUU	UGU	UUG				cysteine, leucine, valine
UUU						phenylalanine
ACG	AGC	CAG	CGA	GAC	GCA	alanine, aspartic acid, serine
ACU	AUC	CAU	CUA	UAC	UCA	asparagine, histidine, threonine
CGU	CUG	GCU	GUC	UCG	UCU	alanine, arginine, serine
AGU	AUG	GUA	GAU	UAG	UGA	aspartic acid, glutamic acid, methionine

tutes of Health. They were able to obtain a cell-free system, from fractionated *E. coli* cells, that would readily combine amino acids to form proteins.

Available theory plus available techniques suggested a critical experiment. What would happen if this cell free system, which could synthesize proteins, was given a synthetic RNA? Could this RNA serve as a messenger?

TABLE 18-2

An example of way genes, messenger RNA, and proteins are thought to be related the data for the RNA codes are from Table 18–1. There are other possibilities than the one given here). The actual mechanism will undoubtedly prove to be more complex than is shown by this simple scheme. For example, there are now some data suggesting that two DNA strands are necessary for messenger RNA to be made.

IF THE SEQUENCE OF BASES IN ONE DNA STRAND IS . . .	. . . THEN THE SEQUENCE OF BASES IN MESSENGER RNA WILL BE . . .	. . . AND THE PROTEIN WILL HAVE THIS SEQUENCE OF AMINO ACIDS
T	A	
G	C	HISTIDINE
G	C	
C	G	
A	U	VALINE
A	U	
T	A	
A	U	LEUCINE
A	U	
T	A	
A	U	LEUCINE
A	U	
G	C	
T	A	THREONINE
G	C	
G	C	
G	C	PROLINE
G	C	
C	G	
A	U	VALINE
A	U	
T	A	
T	A	GLUTAMIC ACID
G	C	
T	A	
T	A	LYSINE
T	A	

Nirenberg and Matthaei added poly U to their system plus an abundance of each of the 20 amino acids. Protein was formed, but *it consisted solely of long chains of the amino acid phenylalanine.* The other 19 amino acids were not joined to make proteins. Thus it seems that uracil alone carries all the information needed to 'tell' the ribosomes to join phenylalanines together. These experiments did not indicate how many uracils were needed but, if the code was a triplet, uracil-uracil-uracil (or UUU) is the code for phenylalanine.

Since August 1961, when these experiments were published, progress has been rapid. Nirenberg and his associates at the National Institutes of Health have tentatively identified the code for each of the 20 amino acids. Another group, associated with Servero Ochoa at the New York University Medical School, has conducted similar experiments. For the most part the results of the two groups agree, as can be seen from the data in Table 18–1, published in February 1963.

These purely tentative results leave many questions unanswered. For example, it has been possible, in only one instance, to determine the order of the bases in the triplet. The sequence for tyrosine appears to be AUU. For the others there are no sure answers. Thus one of the codes for histidine consists of one adenine and two cytosines but it is not known wether the correct sequence is ACC, CCA, or CAC. Some of the amino acids seem to be coded by more than one triplet.

The data and hypothesis of this rapidly advancing field can be summarized as in Table 18–2, where it is shown how a short length of protein may be formed. The example is the section of hemoglobin S that Ingram found to differ from normal hemoglobin (page 163). An understanding of inheritance in molecular terms seems to be at hand.

We are now at the close of our discussion of genetics. In a sense it all began with Mendel. The geneticists following him worked first from the character back to the gene. Later other geneticists worked from the gene, which they discovered to be DNA, to the character. In the process they have given meaning to much of the biochemistry of the cell. The two approaches are now at the point where they can be joined.

We have reached an important stage in the intellectual history of mankind. The accomplishments that began with Griffith's discovery of transformation in Diplococcus and led to the cracking of the genetic code by Nirenberg and Matthaei are without parallel in biology and are difficult to match in any science.

SUGGESTED READINGS

Allison, A. C. 1956. 'Sickle cells and evolution.' *Scientific American 195*:87–94.

Brenner, S., F. Jacob, and M. Meselson. 1961. 'An unstable intermediate carrying information from genes to ribosomes for protein synthesis.' *Nature 190:* 576–81.

Crick F. H. C. 1962. 'The genetic code.' *Scientific American 207*:66–75.

Hurwitz, J. and J. J. Furth. 1962. 'Messenger RNA.' *Scientific American 206:* 41–9.

Ingram, V. 1958. 'How do genes act?' *Scientific American 198*:68–74.

Jacob, F. and J. Monod. 1961. 'On the regulation of gene activity.' *Cold Spring Harbor Symposia on Quantative Biology 26*:193–209.

Nirenberg, M. W. 1963. 'The genetic code: II.' *Scientific American 208*:80–94

Nirenberg, M. W. and J. H. Matthaei. 1961. 'The dependence of cell-free protein synthesis in *E. coli* upon naturally occurring or synthetic polyribonucleotides.' *Proceedings of the National Academy of Science 47*:1588–1602.

Pauling, L., H. A. Itano, S. J. Singer, and I. C. Wells. 1949. 'Sickle cell anemia, a molecular disease.' *Science 110*:543–8.

Watson, J. D. and F. H. C. Crick. 1953. 'Molecular structure of nucleic acid.' *Nature 171*:737–8.

Watson, J. D. and F. H. C. Crick. 1953. 'Genetical implications of the structure of deoxyribonucleic acid.' *Nature 171*:964–7.

Zalokar, M. 1960. 'Sites of protein and ribonucleic acid synthesis in the cell.' *Experimental Cell Research 19*:559–76.

Zamecnik, P. C. 1960 'Historical and current aspects of the problem of protein synthesis.' *Harvey Lectures 54*:256–81.

For a general discussion of the problems raised in this chapter see:

Sager, Ruth and F. J. Ryan. 1961. *Cell Heredity.* John Wiley.

EMBRYOLOGY

PROSPECTUS

Embryology is the branch of biology dealing with the events associated with the formation of an adult individual from a fertilized ovum. The magnitude of the events during this period can be inferred from a consideration of the differences between a fertilized ovum and the adult of an animal such as the frog. The fertilized ovum consists of a single cell, the frog of many billions; and the differences are more profound than mere cell number. A billion frog eggs, no matter how arranged, do not constitute a frog. The cells derived from the single-cell zygote have differentiated along many separate pathways between the fertilized ovum and the adult. Some have formed muscle cells, others neurons, and still others the specialized types found in the liver, stomach, kidney, endocrine glands, and gonads. All the diverse cell types of the adult are derived from a single cell, the fertilized ovum.

There are two main processes involved in development from the fertilized egg to the adult body form. One is a tremendous *increase in cell number*. This increase in cell number is accomplished by mitotic division of already existing cells, or as Virchow expressed it, *omnis cellula e cellula*. The increase in cell number generally involves an increase in the total mass of protoplasm and this, in turn, necessitates the utilization of food from some extraneous source.

The second process is *differentiation*. The term probably suggests the nature of the process. During the course of development, the cells not only increase in number but they change in structure and function. They become different from their earlier forms and from one another.

Our ultimate aim in the study of embryology shall be to ask and attempt to answer some of the pertinent questions concerning development. The 'whys' and 'hows' of development are among the most exciting questions that a biologist can ask. We must begin our study of embryology, however, with a brief description of early development. It is necessary that we have this background before we can ask intelligent questions about the dynamics of the embryological processes. The frog will be used as a type since more is known about its development than about that of any other animal.

19

A Synopsis of Development of the Amphibian Embryo

Meiosis and Fertilization. When the ovum of the frog leaves the ovary, meiosis begins. While the ovum is passing through the body cavity, or when it is in the upper portion of the oviduct, the first meiotic division occurs and the first polar body is formed. Meiosis reaches the stage of the metaphase of the second division by the time the ovum enters the uterus. There are no further nuclear changes until fertilization.

Fertilization occurs as the ova leave the body of the female. A single sperm enters each ovum. The head of the sperm contains the paternal nucleus with its haploid set of 13 chromosomes. Immediately behind the sperm head is a centriole. This will form an essential part of the mitotic apparatus of the embryo's cells.

Meiosis in the egg, which had stopped at metaphase of the second division, is resumed after fertilization. The second polar body is pinched off, leaving the maternal nucleus with the haploid set of 13 chromosomes. The maternal nucleus and the paternal nucleus fuse to form the diploid zygote nucleus with 26 chromosomes. It is this nucleus that divides by mitosis to form all the nuclei of the embryo and adult.

Once fertilization has taken place, the development of the embryo begins. A sequence of definite stages is passed according to a timetable which is dependent on temperature. Our description will be based on *Rana pipiens* embryos developing at 20°C. At 25°, development would take approximately half as long and at 15° nearly twice as long.

The Uncleaved Zygote. The just-fertilized ovum is a sphere approximately 1.7 mm in diameter (Fig. 19–1, this figure and all of those in this chapter show the embryos magnified 25 times). Somewhat more than half of the embryo is a dark chocolate-brown and the remainder is almost white. The center of the dark area is the *animal pole*. It was

Animal Pole

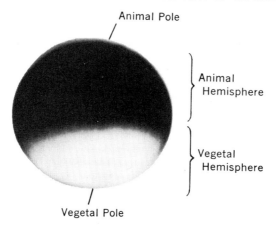

Animal
Hemisphere

Vegetal
Hemisphere

Vegetal Pole

19–1 0-hour embryo. 1-cell stage.

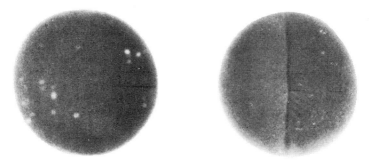

19–2 1-hour embryo. 1-cell stage.   **19–3** 2½-hour embryo. 2-cell stage.

at the animal pole that the polar bodies were formed. The *vegetal pole* is 180° from the animal pole and in the center of the unpigmented area. The *animal hemisphere* is the half of the embryo that has the animal pole in its center. The *vegetal hemisphere* is the half of the embryo that has the vegetal pole in its center.

The entire embryo is surrounded by membranes of a jelly-like substance, which were secreted by the oviduct. (The jelly has been removed from the embryos used in the photographs, and so it cannot be seen.)

Shortly after fertilization, the embryo rotates within its membranes so that the animal hemisphere is uppermost. The orientation that one observes when examining an embryo under a microscope is shown in Fig. 19–2. The animal pole is in the center. Since the pigmented area occupies slightly more than the animal hemisphere, a top view of the embryo shows only the heavily pigmented zone.

 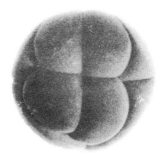

19–4 3½-hour embryo. 4-cell stage.   19–5 4½-hour embryo. 8-cell stage.

The Early Cleavage Stages. Two and one-half hours after fertilization, the first spectacular event in development occurs (Fig. 19–3). A tiny groove appears in the animal hemisphere and this gradually enlarges to form the *first cleavage furrow.* This furrow slowly extends through the embryo until it is divided into two cells. Preceding this external indication of mitosis, the nucleus had gone through the usual prophase, metaphase, anaphase, and telophase stages. The two daughter cells receive a diploid set of chromosomes.

The *second cleavage* occurs about 3½ hours after fertilization (Fig. 19–4). The plane of this cleavage is vertical and at a right angle to the plane of first cleavage. It also begins at the animal pole and extends through the embryo to the vegetal pole. When this cleavage is complete, the embryo consists of four cells.

The *third cleavage* occurs about 4½ hours after fertilization (Fig. 19–5). The plane of this cleavage is perpendicular to the first two. Its position is somewhat above the equator of the embryo, with the result that there are produced four smaller animal-hemisphere cells and four larger cells, which contain the lower part of the animal hemisphere and all of the vegetal hemisphere.

The process of cleavage continues. The embryo becomes divided into smaller and smaller cells (Fig. 19–6). With each division of a cell there is a concomitant division of the nucleus, every daughter cell receiving the diploid set of 26 chromosomes.

The Blastula Stages. The 9-hour embryo (Fig. 19–7) is an early *blastula.* The blastula stage is characterized by an internal cavity, the *blastocoel,* which cannot be seen from the exterior. More will be said about it later when we consider the internal events during early development.

At 14 hours (Fig. 19–8), the embryo is a middle blastula. The rate of mitosis of the cells of the animal hemisphere is more rapid than that

19–6 5½-hour embryo. 16-cell stage.   19–7 9-hour embryo. Early blastula.

19–8 14-hour embryo. Middle blas-   19–9 14-hour embryo. Middle blastula. Top view.   tula. Bottom view.

19–10 22-hour embryo. Late blastula.
Top view.

of the cells of the vegetal hemisphere, so they are now very numerous. If the embryo is turned upside down, the much larger vegetal hemisphere cells are visible (Fig. 19–9).

During the next few hours, continuing cell division is the only visible event. The animal hemisphere cells become so small that they can be distinguished only with difficulty (Fig. 19–10). In this 22-hour late blastula, the next major event of development is foreshadowed. If this embryo is rotated slightly and examined from the side, a special area of pigmentation will be observed (Fig. 19–11). The cells of this pigmented area are in the vegetal hemisphere and slightly below the embryo's equator. It is at this point that the blastopore will form.

Gastrulation. The special pigmented cells noticed at 22 hours gradually become a groove in the surface of the embryo (Fig. 19–12). This groove is the *blastopore* and its appearance marks the beginning of gastrulation. Gastrulation is a process of development that leads to a complete reorganization of the embryo. All of the cells of the area corresponding roughly to the vegetal hemisphere go to the interior of the embryo. The cells of the remainder of the embryo, corresponding roughly to the animal hemisphere, spread and cover the entire embryo. This process of cells turning into the interior is known as *invagination.* The cells are invaginated through the blastopore. The area immediately above the blastopore is known as the *dorsal lip* of the blastopore.

In a few hours the blastopore has changed from a small curved groove to a full semi-circle (Fig. 19–13). Cells are invaginating along the entire length of the blastopore. The overgrowth of the pigmented cells is restricting the area of light-colored cells to a zone of continually decreasing size.

By 30 hours the blastopore is complete (Fig. 19–14). It is a 360° groove into which material is invaginating. The light-colored cells, which are now entirely surrounded by the blastopore, form the *yolk plug.*

The blastopore constricts rapidly and the yolk plug becomes correspondingly smaller as gastrulation proceeds (Fig. 19–15). By 36 hours the yolk plug is very small and the embryo is nearly covered by the overgrowth of cells that were originally restricted to the animal hemisphere (Fig. 19–16). At the end of gastrulation, the yolk plug is drawn into the embryo and the blastopore remains as a tiny slit. At this time the entire outer surface of the embryo is covered by material that was part of the animal hemisphere at the beginning of gastrulation.

The Neurula. The next prominent external change is concerned with the development of the nervous system. Approximately 40 hours after the beginning of development the *neural folds* make their appearance on the top of the embryo (Fig. 19–17). These folds extend, as paired structures, from the blastopore region across the top of the embryo to a point where they join one another. These folds will even-

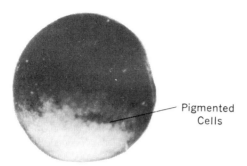

Pigmented
Cells

19–11 22-hour embryo. Late blastula. Side view.

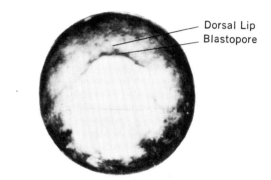

Dorsal Lip
Blastopore

19–12 25-hour embryo. Early gastrula. Bottom view.

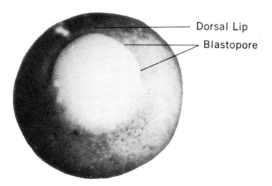

Dorsal Lip
Blastopore

19–13 27-hour embryo. Early gastrula. Bottom view.

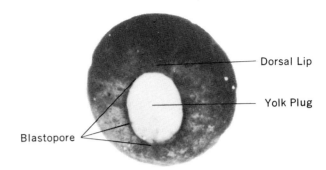

19–14 30-hour embryo. Middle gastrula. Bottom view.

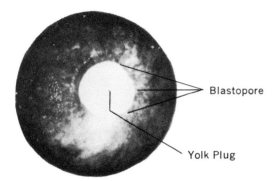

19–15 34-hour embryo. Late gastrula. Bottom view.

19–16 36-hour embryo. Late gastrula. Bottom view.

Anterior

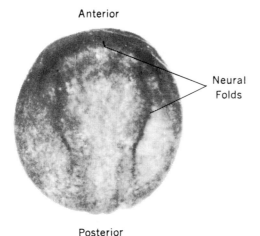

Neural
Folds

Posterior

19–17 42-hour embryo. Early neurula. Dorsal view.

Anterior

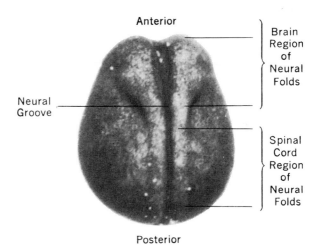

Brain
Region
of
Neural
Folds

Neural
Groove

Spinal
Cord
Region
of
Neural
Folds

Posterior

19–18 47-hour embryo. Mid neurula. Dorsal view.

tually grow together, and in so doing they will form the *neural tube*. The neural tube will develop later into the brain and spinal cord. In the anterior region the neural folds are widely separated. This area will form the brain and the narrower posterior part will form the spinal cord. In Fig. 19–17 the blastopore is not visible, it being below the posterior edge of the embryo.

The growth of the neural folds is a rapid process. By 47 hours the folds are much better developed (Fig. 19–18). The section that will

Anterior

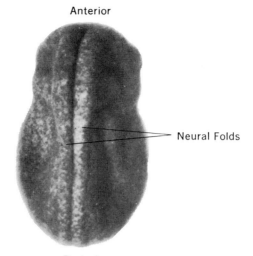

Neural Folds

Posterior

19–19 50-hour embryo. Late neurula. Dorsal view.

Anterior

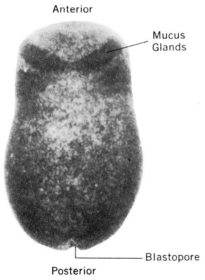

Mucus
Glands

Blastopore

Posterior

19–20 50-hour embryo. Late neurula.
Ventral view.

Anterior

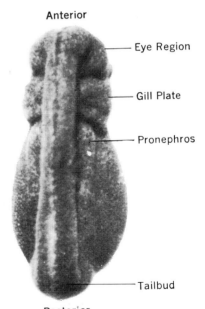

Eye Region

Gill Plate

Pronephros

Tailbud

Posterior

19–21 70-hour embryo. Tailbud.
Dorsal view.

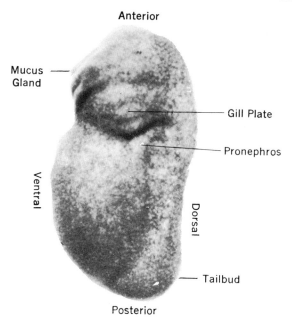

Anterior

Mucus ——
Gland

Gill Plate

Pronephros

Ventral

Dorsal

Tailbud

Posterior

19–22 70-hour embryo. Tailbud. Lateral view.

form the brain and the section that will form the spinal cord are
clearly separated. The area between the folds is the *neural groove*. The
embryo has begun to elongate by this time.

At 50 hours, the two neural folds have come together and the neural
groove closes off as an internal neural tube (Fig. 19–19). On the ventral
side of this same embryo the area that will form the *mucus glands* has
made its appearance (Fig. 19–20). The position of the blastopore is
indicated by a posterior cleft. Later the cloacal opening will form where
the blastopore closed.

   The Tailbud Stage. Figures 19–21 to 19–23 show three views of a
70-hour embryo of the tailbud stage. The dorsal view, Fig. 19–21,
should be compared with the same view of the late neurula (Fig. 19–19).
The neural folds have closed completely and a number of interesting
looking bumps have made their appearance. Near the anterior end
of the embryo there are prominent swellings in which the *eyes* are
forming. Posterior to this is an area that will form the *gills*. Smaller
bumps represent the beginnings of the *pronephros,* which is the embryonic kidney. A tiny tail is forming. In the lateral and ventral view,
the mucus glands are seen as prominent structures. The cloacal open-

Anterior

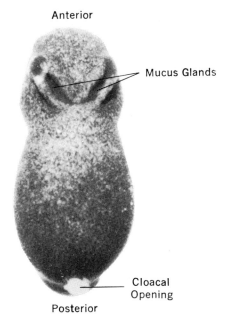

Mucus Glands

Cloacal
Opening

Posterior

19–23 70-hour embryo. Tailbud. Ventral view.

ing (Fig. 19–23) is indicated by a mass of white material that is being extruded.

The 100-hour Embryo. An embryo of 100 hours is the last stage we shall describe (Fig. 19–24). By this time all of the main internal organ systems have begun to form. Externally, the embryo has begun to resemble a tadpole. The eyes are present as bumps on the side of the head. The *olfactory organs* are in the form of pits on either side of the head. The gills have formed, and in a living embryo we would see blood corpuscles streaming through them. The embryo has a well-formed *tail*. It is at approximately this stage that the young tadpole hatches from the jelly membranes, in which it has been encased up to this time. A ventral view of the head (Fig. 19–25) shows the paired olfactory organs, the mucus glands, the gills, and a median depression, the *stomodaeum*. Somewhat later in development the *mouth* forms at the inner end of the stomodaeum.

In a period of four days the single-cell fertilized ovum has become a small tadpole with its organ systems functional. Many—in fact most—of the events that have occurred have been internal. Now that we have some knowledge of the external aspects of development, we can turn to the more complex internal events.

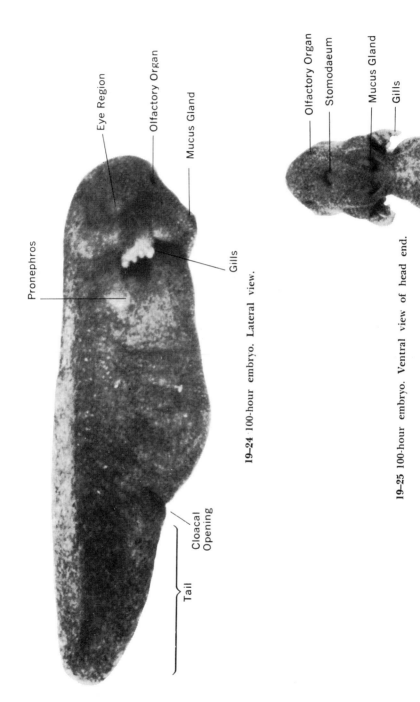

Pronephros

Eye Region

Olfactory Organ

Mucus Gland

Gills

Tail

Cloacal
Opening

19-24 100-hour embryo. Lateral view.

Olfactory Organ

Stomodaeum

Mucus Gland

Gills

19-25 100-hour embryo. Ventral view of head end.

Gastrulation and Organ Formation

In this chapter we shall investigate the processes that convert a single-cell zygote into an embryo having the rudiments of the various organ systems. We shall base the discussion on the development of *Rana pipiens,* which was outlined in the previous chapter.

Structure of the Blastula. During the first day of development, the most obvious process that occurs is mitosis. The mass of the embryo, which was originally in a single cell, is divided into many thousands of smaller cells. The result is the formation of a blastula, which is a spherical embryo with an internal cavity, the *blastocoel* (Fig. 20–1). The blastocoel is restricted to the animal hemisphere, and in the living embryo it contains a fluid. The cells near the animal pole are the

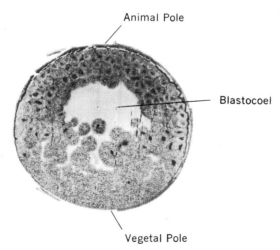

20–1 12-hour embryo. Early blastula. Cross section.

smallest ones in the embryo. Cell size becomes increasingly greater toward the vegetal pole.

All cells of the blastula contain a number of *yolk granules*. These granules were deposited in the ovum while it was being formed in the ovary. They serve as the food supply until the young tadpole is able to feed. The number and size of the yolk granules per cell increase in relation to the nearness of the cell to the vegetal pole—the large cells of the vegetal hemisphere have their cytoplasm packed with yolk granules. Yolk granules are denser than ordinary cytoplasm and, as a result, the vegetal hemisphere, with its abundant supply of yolk granules, is heavier than the animal hemisphere. As a consequence of this density difference the animal hemisphere is uppermost throughout the cleavage and blastula stages.

We have learned previously that gastrulation involves a complete rearrangement of the parts of the embryo. The vegetal hemisphere cells invaginate through the blastopore and the animal hemisphere cells spread over the entire embryo to form its outer covering. These movements are best described in terms of the embryonic structures the cells will form, so it will be necessary to mention some of these. We can then study the cells of the late blastula in terms of their eventual fates.

The Three Embryonic Layers. Throughout the animal kingdom, embryonic development has many features in common. Gastrulation leads to a reorganization in the embryo's structure. At the end of gastrulation, the cells frequently become arranged in three concentric layers, which from the outside to the center are the ectoderm ('outer skin'), mesoderm ('middle skin'), and endoderm ('inner skin').

The *ectoderm* is the outer covering of the late gastrula in the frog. The epidermis and the nervous system develop from the ectoderm.

The *mesoderm* is the second layer of the late gastrula in the frog. It is located under the ectoderm. During the course of development it gives rise to muscles, the skeletal system, the dermis or inner layer of the skin, the circulatory, excretory, and reproductive systems.

The *endoderm* is the inner layer of the late gastrula in the frog. The endoderm gives rise to the inner lining of the alimentary canal and the structures derived from it, such as lungs, the liver, pancreas, and the bladder.

Fate Maps. The cells that form the ectoderm, mesoderm, and endoderm at the close of gastrulation can be located in the earlier stages. It is possible, for example, to map the early gastrula with reference to what its various parts will produce. We might call such a map a 'fate map,' since it will show the developmental fate of the parts of the gastrula.

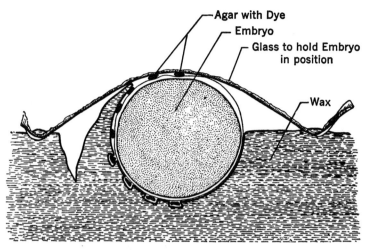

20-2 Vogt's diagram showing how embryos were stained (from *Roux Arch.* 106:565).

The technique for making a fate map of the amphibian gastrula was perfected by the German embryologist Vogt. His method was as follows: Tiny pieces of agar were stained with vital dyes, which are dyes that will not harm living cells (**Fig. 20–2**). The stained agar pieces were then held against the embryo until some of the stain was absorbed by the surface cells. The result was a small colored spot on the embryo. The movement of this colored spot was then traced through development. If the spot was part of an area that invaginated, it was necessary to dissect the embryo to determine where it went.

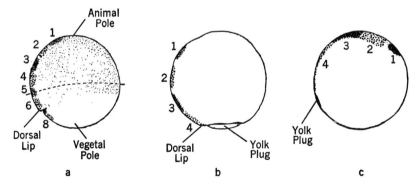

20-3 One of Vogt's experiments with vital stains. *a* is an early gastrula with the dorsal lip below 6. *b* is a mid gastrula. Spots 5, 6, 7, and 8 have been invaginated. The yolk plug is on the ventral side. *c* is a late gastrula. Spots 1, 2, 3, and 4 have spread to cover more of the surface than they occupied at the beginning of gastrulation (from *Roux Arch.* 120:568).

Figure 20–3 is taken from Vogt's work. In the first embryo there are eight zones indicated. These would have been colored in the living embryo, but are represented by dots of different sizes in the diagram. Embryo *a* is an early gastrula with the dorsal lip of the blastopore just formed; *b* is a middle gastrula with a large yolk plug; *c* is a late gastrula with a small yolk plug. The eight colored areas underwent extensive movements during gastrulation. Regions 5, 6, and 8 were all invaginated to the interior of the embryo, and 1, 2, 3, and 4 were stretched to cover a large area of the surface.

After performing numerous experiments of this sort, Vogt was able to prepare a fate map for the European toad, Bombinator (Fig. 20–4). The *presumptive ectoderm,* that is, the area that will form the ectoderm later in development, occupies nearly half of the early gastrula. Two main subdivisions of the presumptive ectoderm are recognized: (1) the *presumptive neural tube,* which will form the brain, spinal cord, nerves, and some of the sense organs, and (2) the *presumptive epidermis,* which forms the outer layer of the skin.

The *presumptive mesoderm* forms a band surrounding the embryo near the equator. It also is divided into two main areas. The *presumptive notochord* is composed of cells that will form the notochord in the neurula. This is a rod of tissue that extends along the dorsal side of the embryo beneath the neural tube. The great importance of the notochord for development will be explained at a later time. The remainder of the presumptive mesoderm will form the other structures derived from this layer, such as the muscular, skeletal, circulatory, reproductive, and excretory systems.

The *presumptive endoderm* is restricted to the ventral portion of the vegetal hemisphere. This area will form the lining of the alimentary canal and structures derived from it, such as the liver, pancreas, and bladder.

The significance of much that has just been said will not be appreciated until we have discussed the later stages. At that time it would be wise to re-read these last few paragraphs.

Gastrulation Movements. During gastrulation all of the material below the line separating the ectoderm and mesoderm is invaginated. The cell movements involved can be understood if we refer to diagrams of sections of progressively older embroyos.

A convenient stage with which to begin is a very late blastula that has the faintest indication of the place where the dorsal lip will appear. An embryo of this stage is shown in Fig. 19–11. A median longitudinal section of such an embryo is shown in Fig. 20–5*a*, and an interpretative diagram is given in Fig. 20–5*b*. The roof of the blastocoel consists of

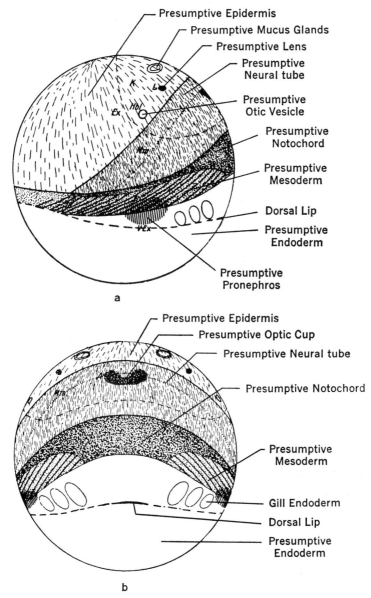

20–4 Vogt's fate map for Bombinator. *a* is a lateral view. *b* is a view toward the dorsal lip (from *Roux Arch.* 120:638).

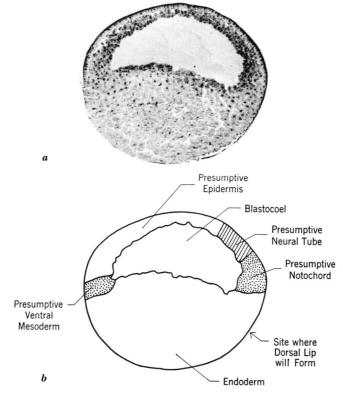

20-5 22-hour embryo. Late blastula. Cross section (*a*) and diagram showing the presumptive regions (*b*).

two regions of ectoderm: the presumptive epidermis and the presumptive neural tube. The portion of the blastocoel roof above the point where the dorsal lip of the blastopore will form is the presumptive notochord. During gastrulation one should pay special attention to the movements of the presumptive notochord and the presumptive neural tube areas.

Gastrulation begins with the formation of the dorsal lip of the blastopore at about 22 hours after fertilization. Figure 20–6 shows a section of a somewhat older 30-hour gastrula. This should be compared to Fig. 19–14 of the whole embryo. Invagination has produced a tiny cavity, the *archenteron* ('primitive gut'). The opening of the archenteron to the outside is the blastopore. The invaginating cells are beginning to obliterate the blastocoel. In this embryo the blastopore is complete, so the ventral lip is seen as a slight invagination. Part of the presumptive notochord cells have turned in to form the roof of

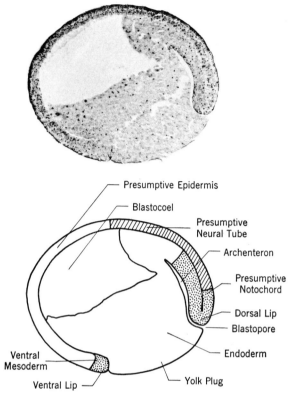

20-6 30-hour embryo. Mid gastrula. Longitudinal section and diagram of the presumptive regions.

the archenteron. These are rolled in over the dorsal lip in a manner analogous to a rope being rolled over a pulley. The presumptive neural tube and presumptive notochord areas are expanding to cover larger portions of the surface.

Four hours later, the gastrula shows important changes (Fig. 20–7, which should be compared with the whole embryo in Fig. 19–15). The archenteron has become larger and there is a corresponding reduction in the blastocoel. (Note that in this figure, as in many others, there are spaces such as the one in the endoderm of the yolk plug. This was not present in the living embryo, but is an artifact resulting from the technique employed in making the preparation.)

In the 36-hour embryo the archenteron has nearly reached its full size (Fig. 20–8, which should be compared with the whole embryo in Fig. 19–16). The embryo is now covered entirely by ectoderm, except for a small amount of endoderm protruding as the yolk plug. All of

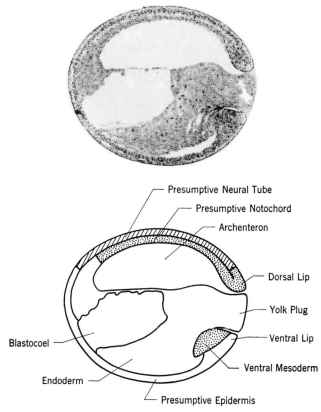

20–7 34-hour embryo. Late gastrula. Longitudinal section and diagram of the presumptive region.

the presumptive notochord cells have invaginated. They form the roof of the archenteron and are situated beneath the portion of the ectoderm that later will form the neural tube. The section shown is not exactly along the mid-line of the embryo which is why the blastocoel does not show. Its position is indicated by dashed lines.

The Neurula. The further history of the presumptive regions will be shown in two older embryos. Figure 20–9 shows a neurula of the same stage as the whole embryo of Fig. 19–18. This is a section along the mid-line, so the neural folds are seen only in the anterior portion of the embryo. The archenteron occupies nearly half of the embryo. It has a ventral outgrowth, the liver diverticulum, which will form the liver. The blastopore does not appear in the section, but its position is indicated in the diagram of the presumptive regions. The yolk plug

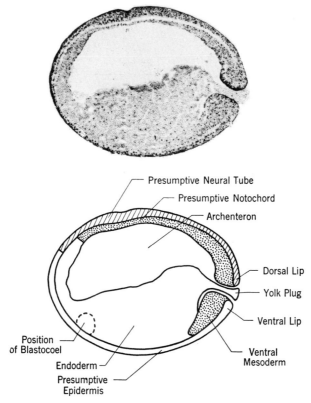

Presumptive Neural Tube

Presumptive Notochord

Archenteron

Dorsal Lip

Yolk Plug

Ventral Lip

Ventral Mesoderm

Position of Blastocoel

Endoderm

Presumptive Epidermis

20-8 36-hour embryo. Late gastrula. Longitudinal section and diagram of the presumptive regions.

has been incorporated in the main mass of endoderm. The blastocoel is but a shadow of its former self.

In the late neurula (Fig. 20–10) the neural tube has formed. The anterior portion is enlarged and it bends ventrally. This is the brain. The neural tube is hollow throughout its length, as shown in the diagram. Only a portion of the cavity of the neural tube is visible, because the section is not exactly longitudinal and median. The epidermis now covers the entire embryo. The mesoderm is represented in this section by the notochord and the ventral mesoderm. The archenteron is now bounded by endoderm on all sides. Earlier, the presumptive notochord formed the archenteron roof, but by the late neurula stage the endoderm has moved up from the sides and formed a layer beneath the notochord.

Some of the details of neural tube formation can be illustrated better

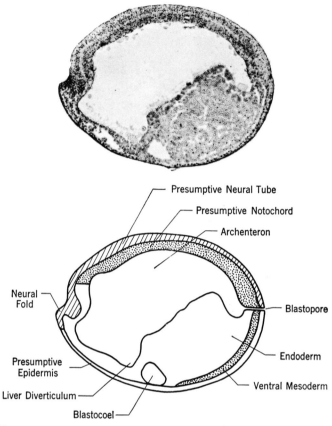

20–9 47-hour embryo. Mid neurula. Longitudinal section and diagram of the presumptive regions.

with cross sections of the embryos. Figure 20–11 shows an embryo shortly before the closure of the neural folds. The ectoderm on the dorsal side has formed two ridges, the neural folds, with the neural groove between them. The remaining surface of the embryo is covered with epidermis, which is also ectoderm. Beneath the ectoderm there is a continuous layer of mesoderm, which forms the notochord on the dorsal mid-line. The remaining portion of the mesoderm is a thin layer, which is difficult to see in the photograph, but is recognizable by the numerous darkly stained nuclei. Most of the embryonic mass is endoderm. This innermost layer forms the thin sides and roof of the archenteron and the thick ventral portion.

The embryo just described shows the characteristic distribution of the three embryonic layers. The outer layer is the ectoderm, beneath

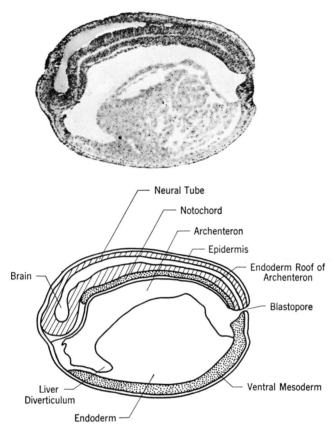

20-10 55-hour embryo. Late neurula. Median longitudinal section and diagram of the parts.

this is the mesoderm, and the innermost one is the endoderm. We might visualize them as three tubes of different sizes fitting into one another.

A few hours later the two neural folds meet and fuse (Fig. 20–12). This forms the neural tube with its central cavity. This cavity extends throughout the neural tube, being especially wide in the brain region. Apart from the closure of the neural folds, this stage is not much different from the preceding one. The embryo consists of the three embryonic layers with little in the way of differentiation.

The 80-hour Embryo. For our purposes it will be necessary to consider one older stage, an embryo of 80 hours. By this time more structures have formed. A section in the anterior end of an embryo of this age shows interesting changes in the neural tube (Fig. 20–13). This

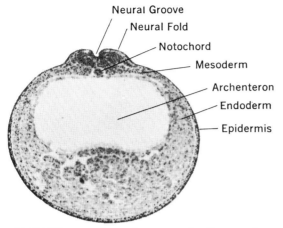

Neural Groove
Neural Fold
Notochord
Mesoderm
Archenteron
Endoderm
Epidermis

20–11 47-hour embryo. Late neurula. Cross section.

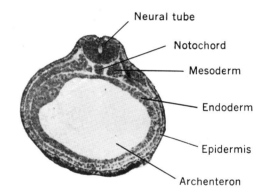

Neural tube
Notochord
Mesoderm
Endoderm
Epidermis
Archenteron

20–12 50-hour embryo. Late neurula. Cross section.

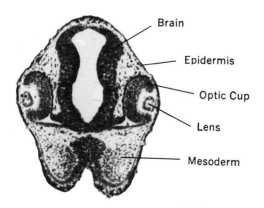

Brain
Epidermis
Optic Cup
Lens
Mesoderm

20–13 80-hour embryo. Cross section in eye region.

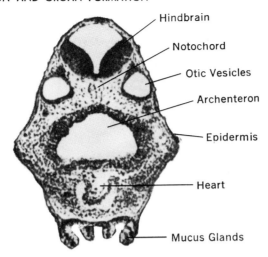

Hindbrain

Notochord

Otic Vesicles

Archenteron

Epidermis

Heart

Mucus Glands

20-14 80-hour embryo. Cross section of heart region.

structure has enlarged to form the brain, and from its ventro-lateral portion the optic cups have grown out. The optic cups will form the retina, which is the portion of the eye that is sensitive to light. The epidermis adjacent to the optic cup forms the lens. Both the optic cup and lens are derived from ectoderm. Note that this section is anterior to both the archenteron and notochord. A study of Fig. 20–10 will show why this is so.

Figure 20–14 is a more posterior section of the same embryo showing other structures. The neural tube at this level is the hindbrain, which is the portion that forms the medulla later in development. The otic vesicles, which were pinched off from the layer of ectoderm covering the embryo, are lateral to the hindbrain. They will be the ears. The archenteron is present in this section, surrounded by its layer of endodermal cells. The heart is forming from mesodermal cells below the archenteron. The epidermis on the ventral side has formed the mucus glands.

A section in the mid-region of the body shows the beginnings of the excretory system (Fig. 20–15). The mesoderm at the sides is forming the pronephros. The mesoderm above the pronephros, which is known as the myotome, will form most of the voluntary muscles of the body. The mesoderm ventral to the pronephric region will split to form a double layer. The space between these two layers is the coelom.

The coelom was mentioned earlier, but not defined. Now we have the background for an adequate definition. A coelom is a body cavity lined by thin membranes (epithelia) derived from mesoderm.

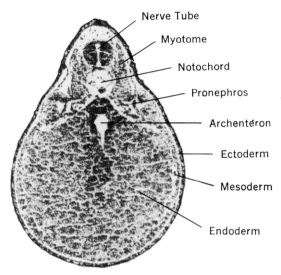

20–15 80-hour embryo. Cross section of mid body region.

This brief survey of early development in the amphibian embryo was designed to provide a background for the consideration of the problems of embryology. Now that we have learned something of *how* an embryo develops we can consider some of the controlling processes that are responsible for embryonic differentiation.

SUGGESTED READINGS

Balinsky, B. I. 1960. *An Introduction to Embryology*. Saunders.

Barth, L. G. 1953. *Embryology,* revised edition. Dryden Press.

Huettner, A. F. 1941. *Fundamentals of Comparative Embryology in the Vertebrates*. Macmillan.

McEwen, R. S. 1949. *A Textbook of Vertebrate Embryology*. Holt.

Rugh, Roberts. 1951. *The Development of the Frog*. Blakiston.

21

Differentiation

The Problem of Differentiation. The main problem of embryology is this: How, in the course of development, does a cell of one type differentiate into one of another type? The problem is all the greater when we realize that the entire structure of the adult is derived from a single cell, the fertilized ovum. The eye, heart, intestine, bone and all the differentiated cells, tissues, and organs can be traced back to this identical beginning.

Up to the present time the data of cytology and genetics are of little help. To the best of our knowledge the genes and chromosomes of the single-cell zygote are the same as those in all adult cells. It is accepted as probable that cellular differentiation is controlled by genes, yet there is no evidence of a differentiation in genes that corresponds to the differentiation in cells.

Differentiation is one of the most important problems confronting biologists today. Some progress has been made in understanding it, but those working in the field are of the opinion that both the data and theories are subject to much doubt.

The problem is not a new one. Aristotle, the first prominent student of embryology, made many observations on developing embryos, and some of his speculations have a surprisingly modern aspect.

In our time one sees progress in nearly all fields of intellectual activity, and one accepts progress as inevitable. It is sobering to realize, therefore, that man's understanding of the factors responsible for differentiation was no better at the time of the American Revolution than it was in the fourth century B.C. In more than two thousand years there was no advance in knowledge on a subject that has interested man throughout his recorded history.

PREFORMATION AND EPIGENESIS

We can begin our study of embryological concepts at the close of this long and sterile period. Much of the speculation of the seventeenth and eighteenth centuries concerned two rival concepts of development, preformation and epigenesis. *Preformation,* as the term implies, means that all the parts of the adult, including the most minute ones, are already perfectly formed at the very beginning of development. *Epigenesis* means that the adult parts are not present at the beginning, but are developed during embryonic life.

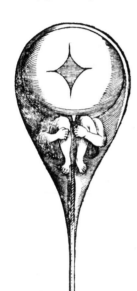

Preformation. Preformation was the generally accepted doctrine during the seventeenth and eighteenth centuries. The basis for this belief was largely philosophical and, to a lesser extent, theological. In addition there are some fascinating 'observations' that seemed to support the concept of preformation.

Some of the greatest scientists of the time, such as Swammerdam, Malpighi, Leuwenhoek, Leibnitz, Réaumur, Spallanzani, and Bonnet, were preformationists. They were divided into two schools: the 'ovists,' who believed that a tiny preformed body was present in the ovum, and the 'spermatists,' who believed that a tiny body was present in the sperm.

Swammerdam, an ovist, believed that the animal hemisphere of the frog's egg contained a tiny frog. This subsisted on the food material of the vegetal hemisphere. Embryonic development was simply the increase in size of this tiny frog. Others thought they saw little chickens in the unincubated hen's egg.

The spermatists reported tiny animals in sperm. Figure 21–1 shows what one observer thought he would find if he could see the inner detail of the human sperm! Many others claimed to confirm him, not only on human sperm but on animal sperm as well.

21–1 Homunculus in human sperm (from Hartsoeker, *Essai de dioptrique,* Paris, 1694, p. 230).

There is no problem of differentiation for the preformationists. The adult structures are already differentiated at the beginning of embryonic life. Development consists only of growth.

The theory of preformation had one interesting corollary. Let us adopt the spermatist position and consider the tiny creature curled up in the sperm head of Fig. 21–1. Let us further suppose that this gametic Lilliputian is a male. If so, his testes must be fully formed and contain sperm. These sperm would contain fully formed creatures as well. They would be the potential children of the homunculus. These children would contain the grandchildren, and the grandchildren would contain the great-grandchildren, all encased like a set of Russian dolls.

Ridiculous as this may seem today, speculations of this sort were made in all seriousness. It was stated by some that Adam or Eve (depending on whether the author was a spermatist or an ovist) must have had the 'seeds' for all future generations of mankind in his or her gonads. The human race would come to an end when the supply of successively encased homunculi was exhausted.

It was probably a philosophical difficulty that made men believe in the theory of preformation when observation might have convinced them of epigenesis. How was one to understand that something entirely new could appear during development? How could the heart, or the brain, suddenly appear where only formless protoplasm existed before? If the egg of the frog was a structureless body detached from any influence of the parent, how was it able to develop into an exact replica of its species? Why did it not form a toad, or a fish, or an elephant? Clearly something 'preformed' must be transmitted from parent to offspring, or the frog's egg would not develop into a frog. No doubt it was easier for the embryologists to imagine that structure, rather than anything else, was transmitted. For this reason it was assumed that the earliest embryo must be a miniature adult.

As an example of how reason can lead one to deny what his senses reveal, we should say a few words about Marcello Malpighi, an outstanding seventeenth-century Italian biologist. (Malpighi made many discoveries. He was, for example, the first to observe capillary connections between arteries and veins. Harvey, who is generally credited with the theory that blood circulates, assumed that connections must exist, but he did not see them.) Malpighi was a preformationist, yet his work on the hen's egg showed that development was epigenetic. The figures in 21–2 are copies of his drawings of the developing chick. In the unincubated egg there is a tiny embryo. (In the chicken, fertilization is internal; some development occurs before incubation begins.)

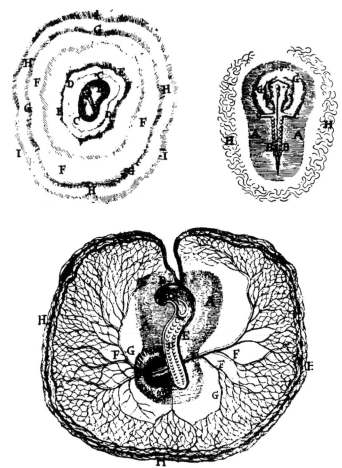

21-2 (legend on facing page).

In increasingly older embryos new structures made their appearance, and gradually a chicken-like form is produced. Malpighi believed that this epigenetic phenomenon was an illusion. To him, the structures that made their appearance were there all the time; he was just not able to see them. If Malpighi had not let his beliefs overrule his observations, posterity would have hailed him for presenting the first well-documented case for epigenesis.

Malpighi should not be criticized for disbelieving what he saw. Scientists throughout the ages have had to face conflicts between preconceived beliefs and observations. Not infrequently the observations are made to fit the preconceived beliefs. Sometimes this has been proved, by subsequent events, to be the correct thing to have done.

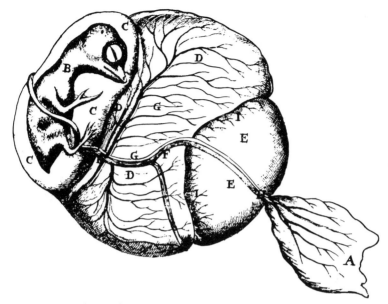

21-2 Four stages in the development of the chick (from Malpighi, 'De Formatione pulli in ovo,' in *Opera Omnia,* Scott and Wells, London, 1686).

For example, consider the movement of the sun and earth. Our senses tell us that the sun moves around the earth daily. Why do you doubt that it does? Another example is Flemming's belief, mentioned in Chapter 2, that chromosomes are constant cell structures. He held this view in spite of the fact that the chromosomes seemed to disappear between successive divisions. At other times the acceptance of beliefs contrary to observation has retarded the progress of science for years. In Malpighi's case, observation should have triumphed over reason. It did not. We must realize, of course, that a verdict of this sort can be given only in retrospect. It is not intellectually dishonest to fit ideas to observations but it is an almost compulsive orientation of the mental processes seeking avidly to fit observations into reasoned order.

Epigenesis. In spite of an almost universal belief in preformation, there were always men of renown who thought that development was epigenetic. They believed that adult structures were absent from the early embryo, and that they made their appearance during the course of embryonic life. This was the belief held by Aristotle and, two millennia later, by the English biologist Harvey (1691).

In 1759 the dissertation of Wolff, the German zoologist, was published. This was the most careful description of the developing chick that had been made. Wolff believed in epigenesis, and the observations reported in his dissertation appeared to confirm his belief, but during his lifetime he was unable to convince many of his contemporaries. In fact it was not until the early part of the nineteenth century that a majority of biologists finally accepted the concept. Its final acceptance resulted from the careful observations of Wolff and those who followed him, who showed that new structures do make their appearance in development.

Epigenesis is, of course, the view that we hold today. The study of the photographs of frog embryos (Figs. 19–1 to 19–25) will indicate that the fertilized ovum is not an adult in miniature. Development begins in a relatively structureless and homogeneous mass, and gradually the organs and parts differentiate to produce the adult body. The problem of differentiation, which would be non-existent if preformation were the case, returns in full force and must be explained.

MOSAIC AND REGULATIVE DEVELOPMENT

The Mosaic Theory of Development. The next major attempt to solve the problem of differentiation was made by the German embryologist Wilhelm Roux in the last two decades of the nineteenth century. He was working at a time when spectacular discoveries of chromosome movements in mitosis and meiosis were being made. Roux attempted to apply the cytological information to embryological problems. According to him, the zygote nucleus contains the *determinants* for differentiation. These determinants were localized in the chromosomes, and during cleavage there was a segregation of determinants. Finally, each cell would be left with only the determinants of a single sort, such as heart cell determinant, muscle cell determinant, and so forth. The theoretical biologist, Weismann, also had a prominent role in the development of this theory.

Cleavage and the Polarity of the Frog Embryo. This hypothesis was suggested by Roux's observation that the plane of first cleavage of the frog's egg coincides with the long axis of the embryo. He noted further that shortly after fertilization and before first cleavage a *gray crescent* formed. This was an area, near the equator of the embryo, from which much of the black pigment disappeared. Later on in development, the dorsal lip of the blastopore begins to form in the area where the gray crescent had been. During gastrulation the position of the blastopore shifts and later the anus forms very near the region where the blastopore closes in the neurula stage. The gray crescent, dorsal lip, and anus are all median structures so far as the embryo is concerned. The plane of

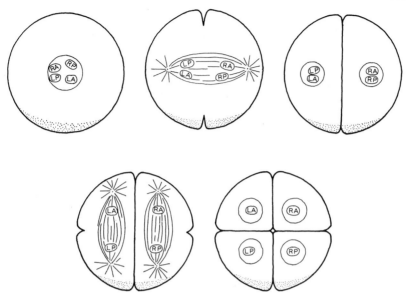

21-3 Roux's theory of the segregation of determinants during cleavage. *LA* is the determinant for the left anterior quadrant of the embryo. *RA, LP,* and *RP* are determinants for the other quadrants, Gray crescent area stippled.

first cleavage, according to Roux, cuts through the middle of the gray crescent so one of the cells will represent the right side of the future embryo and the other the left side.

Roux believed that the first mitotic division of the embryo resulted in the determinants for all structures on the right side going into the right cell and those for the left side going into the left cell. The plane of the second division is at right angles to the first. It cuts across the body of the future embryo. The four cells will represent the right anterior, left anterior, right posterior, and left posterior sections of the future embryo. At the second division, Roux believed, the determinants for these four regions of the future embryo are segregated. The simple diagram of Fig. 21-3 will show how the determinants were thought to be allocated to each cell by the mitotic division. Continuing cell division results in a continuing segregation of determinants. Finally, each cell will contain a specific determinant, which will be responsible for its differentiation into a specific type of cell.

Roux's Test of the Mosaic Theory. Roux sought to test this hypothesis in one of the first experiments ever performed on an embryo. If the hypothesis is true, one of the cells of the two-cell stage should contain the determinants for the left side of the body, and the other cell

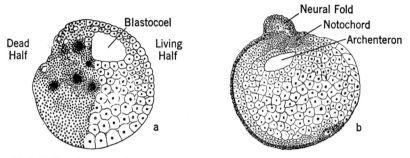

21–4 Half-embryos obtained after killing one cell of the 2-cell stage. *a* is a blastula with the dead cell adjacent to the living half-embryo. *b* is an early neurula. The dead cell has been cast off (from Roux, 1888. *Virchow's Archiv.* 114:113).

the determinants of the right side of the body. Now, if one of the cells is killed, the embryo will have the determinants for only one side of the body. It should develop into a half-embryo. Roux performed this experiment by killing one of the two cells with a hot needle. The uninjured cell cleaved as though it were still part of the entire embryo. It formed a half-blastula (Fig. 21–4), underwent an abnormal gastrulation, and produced a structure that was interpreted as a half-embryo. A single neural fold was formed, and the mesoderm was present only on the uninjured side.

Roux's experiment seemed to be a dramatic proof of his theory of the segregation of determinants during cleavage. The uninjured cell formed only that part of the embryo it would have formed in an uninjured embryo. Roux interpreted this to mean that each cell was capable of forming only the structures it did in normal development. One could imagine the embryo to be a mosaic of cells, each cell capable of producing only a specific part of the adult body.

Mosaic Development in Other Embryos. Other investigators, experimenting with embryos of many kinds of animals, obtained results that seemed to confirm Roux's belief. In some annelid worms it was found that cleavage is a very exact phenomenon. All the eggs cleave in the same manner, and it is possible to trace the history of single cells in development. For example, it was observed that one cell, which forms at the sixth cleavage, gives rise to all of the mesoderm. In some species of tunicates (marine animals related to the vertebrates) the eggs had a pronounced pigment pattern. This enabled the observer to trace the various regions during development, in much the same way as Vogt did many years later in his experimentally stained amphibian embryos. It was found that each region of the early embryo gave rise to specific structures in later embryos.

Embryos of annelid worms, molluscs, tunicates, and several other groups were found to behave in this manner. They were spoken of as exhibiting *mosaic development*. The embryo was regarded as an association of independent cells each developing along a predestined path to form a specific part of the adult.

The Reinterpretation of Roux's Results. The novelty of Roux's theory and its experimental verification convinced many embryologists that an understanding of differentiation was at hand. It was not long, however, before complications and exceptions were noticed. Roux's experiment to test his theory was found to be inadequate. Studies of sea-urchin embryos showed that not all development was of the mosaic type.

It was found that the results Roux obtained after killing one cell of the two-cell stage embryo were due largely to the presence of the dead cell. Whether for mechanical or other reasons, the dead cell seemed to prevent the normal cell from rounding up and producing a whole embryo. If the experiment was done in a different way, it was found that a single cell of the two-cell embryo could form an entire embryo. Thus, if one of the cells was removed by sucking it out with a pipette, the remaining one produced an entire embryo, which differed from the normal only in being smaller. Another technique gave similar results. If a fine thread was tied around the embryo at the two-cell stage, it could be tightened until the two cells were separated. Each of the cells then produced a normal embryo. One could conclude from these experiments that there was no evidence of a segregation of determinants as postulated in Roux's theory.

Regulative Development. In the embryo of the sea urchin, it was found that development is not mosaic. The cells of the two-cell stage could be separated and each cell would give rise to a normal larva. The same was true at the four-cell stage; each cell, if isolated from the rest, could form an entire embryo. In contrast to the mosaic embryos, in which each cell could form only the part it would in normal development, the sea-urchin embryos were said to be *regulative*. Isolated cells of the regulative eggs could adjust to the new situation and produce a whole embryo.

The conclusion was reached that some early embryos are mosaic and some regulative. The significance of the difference will not be apparent until we have discussed the organizer concept, so discussion of the question must be postponed.

With respect to Roux's theory of segregation of nuclear determinants during mitosis, the evidence subsequently obtained indicated that the nuclei of all cells are identical. There were no data to suggest that there existed specific determinants for the many cell types of the adult organism. None is available even today, so the problem of the mode of the nuclear control of differentiation is unsolved. For many biologists this is the most pressing problem requiring a solution.

THE ORGANIZER THEORY

The next major advance in the study of differentiation was made by the German embryologist Hans Spemann and others in experiments on the differentiation of the amphibian nervous system. Their studies led to the organizer theory, now to be considered, which is the most important embryological concept proposed during the first half of the twentieth century.

Formation of the Neural Tube in the Amphibian Embryo. If we examine the cells of an amphibian embryo in the late blastula stage we find that they are essentially the same in appearance throughout the entire embryo. There is a gradient of increasing cell size extending from the animal to the vegetal pole, and the concentration of yolk granules in the individual cells is subject to variation, but beyond this there is little to suggest the widely divergent destinies that will befall the cells of different regions. The conversion of the single-celled zygote into a many-celled blastula is brought about by cleavage, with little or no visible differentiation of the cells: the cells just get smaller. During gastrulation the cells become rearranged and the three germ layers can be distinguished, but even at this time there is little difference among the ectoderm, mesoderm, or endoderm cells. Subsequently the slow process of cellular differentiation results in various visibly different cell types, such as muscle, gland, and nerve, that make up the tissues and organs of the embryo.

We have previously learned that the nervous system is one of the first organ systems to make its appearance in the course of development. Observations of the living embryo give us considerable information concerning the formation of this system. We notice that at the end of gastrulation the embryo becomes flattened on the dorsal side. This flattened area is the neural plate. Next the neural folds appear as ridges along the periphery of the neural plate. These folds move toward the mid-line and fuse along their crests. In this manner the neural plate is converted into a tube lying beneath the now continuous ectoderm in the dorsal part of the embryo.

Observations of the living embryo could be supplemented by the study of sections prepared by the usual histological techniques. From these we could obtain information on the changes occurring within the embryo. We would find that by the time the neural plate is formed, gastrulation movements have brought a sheet of mesodermal cells into position beneath the neural plate. Somewhat later these mesodermal cells will form the notochord and myotomes.

Repeated observations would show that the neural plate and tube are always formed in the same part of the embryo (Fig. 20-4). In normal development, the ectoderm cells situated on the side of the embryo above the blastopore and their descendants produce these structures. The appearance of the neural plate in a constant position suggests that the cells which form it differ in some way from other ectodermal cells. They alone develop into the neural plate, while the remaining ectoderm cells produce the epidermal covering of the body. It is possible to trace the positions of the presumptive neural plate and the presumptive epidermis cells back to the early gastrula, as was done by Vogt (Chapter 2). In all probability, we could even trace the presumptive regions back to the early cleavage stages.

Hypotheses of Neural Tube Formation. What is different about the portion of the ectoderm that will form the neural tube? Why does it and no other part of the ectoderm form this structure? Observations of a normally developing embryo cannot answer questions of this kind. We can only attack the problem by experimentation. Yet what experiment can we perform? First we must formulate a question that is definite enough to have a definite answer for us to seek. We already know that the group of cells that occupies the presumptive neural tube region of the early gastrula will, in later stages, form the neural plate and, still later, the neural tube. Two alternative hypotheses to explain this phenomenon could be suggested:

Hypothesis 1. The presumptive neural tube cells of an early gastrula possess an inherent capacity to form neural tissue. That is, they have within themselves all that is necessary to differentiate into a neural tube.

Hypothesis 2. The presumptive neural tube cells of an early gastrula do not possess an inherent capacity to form neural tissue. Influences from outside the presumptive neural tube area are necessary for differentiation of a neural tube.

Tests of the Hypotheses. These hypotheses are formulated in such a manner that they can be tested. Let us begin with the first hypothesis. If the presumptive neural tube cells possess within themselves all that is necessary for neural tube differentiation, we can make the following deduction: The presumptive neural tube cells should be able to differentiate into a neural tube if they are separated from the remainder of the embryo. This separation can be accomplished in several ways.

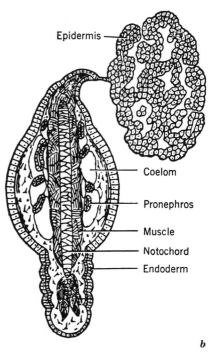

Epidermis

Coelom

Pronephros

Muscle

Notochord

Endoderm

b

21-5 Holtfreter's exogastrulation experiment. *a* is an exogastrula and *b* is a diagram of the parts (from Holtfreter, 1933. *Biologischen Zentralblatt,*
a   53:404).

Experiment 1. Separation of the presumptive ectoderm from the remainder of the embryo in exogastrulation. This experiment was performed by Holtfreter. He removed the membranes from an early gastrula, oriented it with the animal hemisphere down, and let it develop in a solution with a salt concentration higher than in pond water. Under these conditions gastrulation movements were quite abnormal. The presumptive ectoderm cells did not move down over the vegetal hemisphere, but tended to pull away from the remainder of the embryo (Fig. 21-5). This resulted in a dumbbell-shaped embryo known as an *exogastrula.* In extreme cases the presumptive ectoderm cells formed a ball connected by only a thin strand of cells with the presumptive endoderm and mesoderm.

Further development of the two parts was very different. The cells of the presumptive endoderm and mesoderm were able to differentiate into a heart, muscles, parts of the alimentary canal and other organs normally formed from these two layers. In marked contrast, the presumptive ectoderm cells remained undifferentiated. No trace of a nervous system was found. This indicates that the presumptive neural tube cells do not possess an inherent capacity to form a neural tube. If this is so, then Hypothesis 1 is false. We must not be too quick to accept this conclusion, however. Perhaps the manipulation of the embryo injured the presumptive neural tube cells in such a way as to prevent them from differentiating into neural tissue. If this was the case, the experiment is of no value for our purposes. This explanation is improbable, however, since the presumptive endoderm and mesoderm showed a considerable amount of differentiation after being subjected to the same experimental conditions.

Experiment 2. Explantation of presumptive neural tube cells. This experiment has been performed by Holtfreter and several other investigators. Pieces of the blastocoel roof of an early gastrula can be cut off and raised in dilute salt solutions (Fig. 21–6). They remain alive for many days. This process is known as *explantation* and the pieces as *explants*. Explants from the presumptive neural tube and presumptive

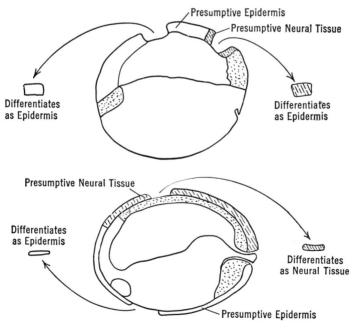

21–6 Explantation of presumptive epidermis and presumptive neural tissue in early (above) and late (below) gastrulae. (Refer to figs. 20–5 and 20–8 for full labels.)

epidermis areas of an early gastrula fail to differentiate into neural tissue. They produce nothing more than a simple epidermal type of cell.

Results of Experiment 2 likewise indicate that Hypothesis 1 is false. We can only hope that neither the cutting nor the culture conditions injured the explant in some manner that would prevent neural differentiation. Such a possibility can partially be ruled out, since explants from some other parts of the embryo are able to differentiate.

If this experiment is repeated at the end of gastrulation the results are different (Fig. 21–6). Explants of presumptive epidermis form only epidermis but explants of presumptive neural tube cells differentiate into neural tubes. Some important change has occurred in the presumptive neural tube cells during the interval between the beginning and end of gastrulation.

The two experiments have given similar answers. If the presumptive neural tube cells are removed from an early gastrula, either by exogastrulation or explantation, no neural tissue is formed. If these results can be accepted, and it seems probable that they can, the presumptive neural tube cells in an early gastrula do not possess an inherent ability to form neural tissue. Thus, Hypothesis 1 is false.

One part of Experiment 2 did show that the presumptive neural tube cells of the late gastrula possess an ability to differentiate into neural tissue if explanted. Clearly, some change has occurred in the interval between the early gastrula stage and the stage immediately before the neural plate forms. This change did not occur during the development of the ectoderm in exogastrulae or in explants. Thus, it is likely that normally an influence from some non-ectodermal part of the embryo is responsible for this change that occurs in the presumptive neural tube cells. This conception is our second hypothesis, which was stated thus: 'The presumptive neural plate cells of an early gastrula do not possess an inherent capacity to form neural tissue. Influences from outside the presumptive neural plate area are necessary for differentiation.' One deduction that we might make from this second hypothesis is that the neural tube should form in the same position, relative to the non-ectodermal parts of the embryo, no matter how the presumptive ectoderm is oriented. This can be tested as follows:

Experiment 3. Rotation of the animal hemisphere of an early gastrula. This experiment was performed by Spemann. It will be recalled that the presumptive neural tube cells occupy that portion of the animal hemisphere nearest to the dorsal lip (Fig. 20–4). The presumptive epidermis cells are on the opposite side of the embryo. If the upper portion of an early gastrula is cut off, rotated 180°, and put back, the cells will come into new relations with the remainder of the

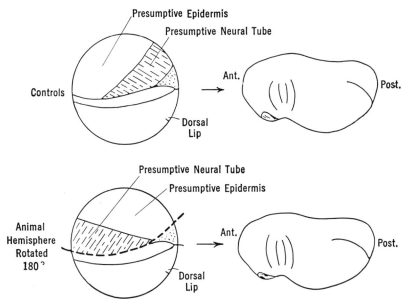

21-7   Rotation of the animal hemisphere. (Refer to fig. 20–4 for full labels.)

embryo (Fig. 21–7). The presumptive epidermis will now be closer to the dorsal lip and the presumptive neural tube will be on the far side of the embryo.

The development of the embryo on which the operation was performed continues as though nothing has happened. The neural folds are formed in the normal relation to the blastopore. *This means that the neural folds of this experimental embryo are formed largely from presumptive epidermis, and that the epidermis is derived almost entirely from the presumptive neural tube cells.*

The results of Experiment 3 indicate that Hypothesis 2 may be correct. The differentiation of the ectoderm appears to be greatly influenced by the ventral part of the embryo. Furthermore, the constant relation with the blastopore suggested to Spemann that the cells invaginated at the dorsal lip might be the stimulus for neural differentiation. In normal development these cells form the roof of the archenteron. Thus, they are in a position immediately beneath the cells that will form the neural tube. If we regard this relation as significant, we might restate the hypothesis in a more specific manner, such as, 'The presumptive neural plate cells of an early gastrula do not possess an inherent capacity to form neural tissue. Differentiation into neural tissue is the result of stimulation by the roof of the archenteron.'

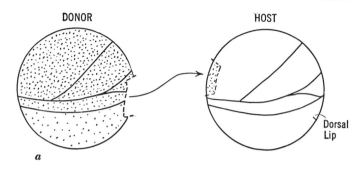

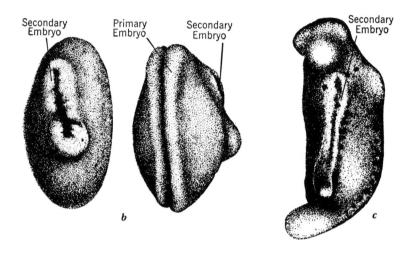

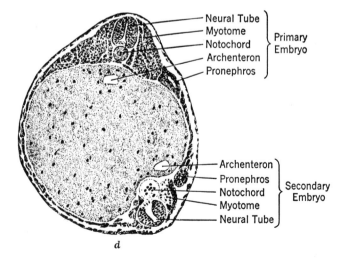

If this hypothesis is true, we might make the following deduction: If the dorsal lip cells are removed from one embryo and grafted onto another, and if they are able to invaginate under the ectodermal cells in this new position, we would expect these ectodermal cells to form neural tissue. It has proved possible to test this deduction by an extremely delicate micro-surgical experiment.

Experiment 4. Transplantation of the dorsal lip. This experiment, reported by Spemann and Mangold in 1924, is one of the most important in embryology. Since it was in actual time the first to be done, it forms the basis of the organizer theory. The embryos of two species of salamanders were used. In one species the embryos were nearly white and in the other they were brownish. A small piece of tissue was removed from the dorsal lip region of one embryo (Fig. 21–8). This piece of tissue was then transplanted to an early gastrula of the other species. The embryo from which a dorsal lip is removed is called the *donor*. The embryo to which the transplant is made is called the *host*. Since the host and donor tissues were of different colors they could be distinguished even when present in the same embryo.

Spemann and Mangold's operation did not seem to affect the process of gastrulation in the host. Invagination occurred at the host's blastopore. Of greater significance was the fact that invagination occurred where the transplanted dorsal lip was placed. The donor cells invaginated through this secondary blastopore and produced a tiny archenteron. At the time the host's neural folds were forming, neural folds also appeared above the region where the donor tissue had invaginated. These neural folds were composed of host cells. The transplanted dorsal lip had exerted a profound influence. Host cells, which in the course of normal development would have produced epidermis, were affected in such a way that they formed neural folds. Later these neural folds closed. When the embryos were examined in sections it was found that the transplanted dorsal lip had stimulated the formation not only of a nerve tube but other structures as well. In some cases almost an entire normal embryo arose in the region where the transplant was made.

Interpretation of the Results. Spemann and Mangold, recognizing that the dorsal lip was of great importance in the formation of em-

---

21–8 Dorsal Lip Transplantation experiment of Spemann and Mangold. *a* is a diagram of the operation. *b* is a lateral and dorsal view of a neurula with the secondary embryo. *c* is an older stage. *d* is a cross section showing the structures of the primary and the secondary embryos. (*b, c,* and *d* from Spemann and Mangold, *Archiv für Mikroskopische Anatomie und Entwicklungsmechanik*, 100:599. 1924.)

bryonic structure, called it the *organizer*. This organizer was postulated to act on other cells and alter their course in development. This action is spoken of as *induction*.

This experiment, and many others, has shown that the neural tube of a normal embryo is formed under the influence of the organizer. At the beginning of gastrulation, the organizer region consists of cells above the dorsal lip, in the area corresponding roughly to the presumptive notochord of Vogt's map (Fig. 20–4). This region invaginates to form the roof of the archenteron. The roof of the archenteron then induces the overlying ectoderm cells to form a neural tube.

We now have the theoretical basis to interpret the results of the experiments on explanting pieces of presumptive neural tube cells. When these cells are explanted at the early gastrula stage they form epidermis but no neural tubes. When the explants are removed at the end of gastrulation they are capable of forming neural tubes. The explanted early-gastrula presumptive neural tube cells did not form neural tubes because the organizer had not acted upon them. At the end of gastrulation, on the other hand, the organizer area would be in the archenteron roof and the overlying presumptive neural cells would have been induced.

The early-gastrula presumptive neural tube cells are said to be *undetermined* so far as the ability to form a neural tube is concerned. Once the organizer has acted upon them they are *determined*. Said in another way, determination results from induction. After determination has occurred differentiation is possible. Both induction and determination are invisible and are presumably biochemical events. They are tested by explantation and other experimental techniques. If a piece of tissue is explanted in a suitable medium and it does not differentiate in a specific way, it is said to be undetermined. If it differentiates after explantation, it was already determined at the time it was removed from the embryo.

These various experiments have enabled us to choose between the two hypotheses concerning the formation of a neural tube, namely, whether (1) the presumptive neural tube cells of an early gastrula have an inherent ability to form a neural tube, or whether (2) these cells must be acted on by other parts of the embryo in order for them to differentiate in their specific manner. All the evidence given indicates that the second hypothesis is correct. At the beginning of gastrulation, the presumptive neural tube cells do not possess an inherent ability to form a neural tube. The only inherent ability they possess is to form simple epidermal cells. The ability to form a neural tube results from the induction of the presumptive neural tube cells by the organizer in the archenteron roof. At the beginning of gastrulation the entire presumptive ectoderm, whether it is in the presumptive neural tube or presumptive epidermis region, is identical in developmental capabilities. One portion forms a neural tube because the organizer acts on it; the remainder forms epidermis because the organizer does not come in contact with it.

It was soon found that there is not just one organizer but, in fact, many organizers exist. Organizers are known for the mouth, heart, lens, otic vesicle, pronephros, and for other structures as well. We shall consider one example of these *secondary organizers,* so called because they originate in structures which themselves have been formed under the influence of the primary organizer of the archenteron roof.

Formation of the Lens. Shortly after closure of the neural folds, paired outgrowths appear in the ventro-lateral portion of the brain. These are the optic cups. They grow laterally until they reach the epidermis. The epidermal cells adjacent to the optic cups then differentiate into the lenses (Fig. 20–13). Experiments using embryos stained with vital dyes show that the cells that will form the optic cup and the cells that will form the lens occupy very different regions of the early gastrula. The former are located in the presumptive neural plate, while the latter are in the presumptive epidermis. The two areas are brought into close association by the complex movements of gastrulation and neural tube formation. Experiments have shown that, in some species at least, the optic cup contains the inductor which is the stimulus for the formation of the lens.

Let us experiment with an embryo in which the optic cups are beginning to form by making a slit in the epidermis over the brain. The optic cup on one side can be cut off and the epidermis put back in place. The cut tissues heal in a matter of minutes, and the embryo appears to suffer no injury as a result of the operation. We then allow the embryo to develop. The optic cup on the unoperated side grows out toward the epidermis and the epidermis produces a lens. On the operated side there is no formation of an optic cup after the first one was removed and *no lens is formed by the presumptive lens area of the epidermis.* In the absence of an optic cup, lens differentiation does not occur. This result suggests that the optic cup may be acting as an organizer for lens formation.

Another type of experiment substantiates this hypothesis: Let us remove the optic cup from an embryo of the proper age. Next we make a slit in the epidermis of the flank region of this same embryo and push the optic cup into this slit. This slit will heal and the optic cup will then be adjacent to epidermis that would normally form only the outer layer of skin. Under the influence of the optic cup, however, the fate of this epidermis is changed—it differentiates into a lens. The embryo then has a structurally normal eye in the flank. (It is non-functional, however, since the proper nerve connections with the brain are not made.)

Obviously, these results are similar to the archenteron roof-ectoderm relationships. In the present case the optic cup is an organizer. It induces lens formation in any ectoderm of the right age with which it comes into contact. In normal development, this is the ectoderm of the head but, as experiments have demonstrated, any ectoderm will suffice.

The Nature of the Organizer. Following the discovery of the neural tube organizer, many embryologists sought to learn as much as they could about it.

The location of early gastrula cells that were capable of induction was found to be restricted roughly to the presumptive notochord and adjacent mesoderm. Transplants of living cells from other parts of the early gastrula did not induce neural structures.

The most encouraging discovery was the ability of dead organizer tissue to induce. Pieces of the dorsal lip, or of the archenteron roof, could be killed by heat or chemical means and still stimulate the formation of neural tissue in undetermined ectoderm. This suggested that the organizer was a chemical substance, and its stability after heat or chemical treatment was a sign that it might be extractable in a pure form.

This early optimism was soon shattered by discoveries that were both confusing and difficult to interpret. The organizer was found to be much more widely distributed than the early experiments on living tissue had indicated. Parts of the gastrula which would not induce neural tissue when alive were found to induce after being killed. Even more perplexing was the finding that adult tissues, such as liver and kidney, could induce embryonic structures. In addition, some organic compounds appear to induce.

Some investigators have attempted to purify the organizer, but the results so far obtained have not been very convincing. Some recent experiments suggest that the primary organizer is nucleic acid.

Further advances in our knowledge of the nature of the organizer are awaiting new ideas and new techniques.

The Reacting Tissue. In our discussion of induction we have emphasized the organizer. This may have left the reader with the impression that the tissue which reacts to the organizer is a passive material that is merely molded by the organizer. This is not the case.

The ability of tissue to respond to organizers is limited in a number of ways. Some of these will now be mentioned.

We have seen earlier that any portion of the presumptive ectoderm can respond to the archenteron-roof organizer by producing a neural tube. The period during which this response is possible is very short. At or about the stage when the neural folds close, the presumptive epidermis will no longer respond to the archenteron-roof organizer. The ectoderm also has a specific period during which it can respond to the optic-cup organizer and produce a lens.

The importance of the reacting tissue can be shown in experiments involving tissue of two different species. The mouth region of a frog larva and that of a salamander larva differ considerably. The frog larval mouth is bordered by prominent black, horny 'jaws' and rows of tiny 'teeth.' These jaws and teeth are ectodermal structures that have no relation to the jaws and teeth of the adult. The salamander larva lacks both the ectodermal jaws and teeth. The anterior portion of the archenteron of both frog and salamander induces the mouth region. An interesting experiment can be performed by interchanging frog and salamander ectoderm in the region where the mouth will form. The salamander embryo will then have its mouth region covered with frog ectoderm. The frog embryo will have its mouth region covered with salamander ectoderm. Which type of mouth will form in the two cases?

The results of such experiments are clear-cut. The tissue always responds in accordance with its specific genetic constitution. The frog tissue on the salamander embryo is induced by the salamander mouth organizer to form a mouth. The mouth it forms, however, is of the frog type. In the same way, the salamander ectoderm on the frog embryo produces a salamander mouth.

A somewhat similar situation is encountered with respect to two structures in the head region of frog and salamander embryos. The frog has a pair of mucus glands on the ventral side near the mouth (Figs. 19–24, 19–25, and 20–14). These are ectodermal structures and they are induced by the underlying tissue. Salamander embryos lack these mucus glands. Instead, they have a pair of balancers (Fig. 21–9). These balancers are also induced by the underlying tissues. If presumptive epidermis of a frog embryo is transplanted to the region behind the mouth in a salamander embryo, the transplant forms mucus glands. The embryo shown in Figure 21–9 is the result of an operation of this sort. The stringy material, which appears below the host's balancer, is mucus being secreted by the induced mucus glands.

Secretion
From Induced
Mucus Glands

21-9 Induction of mucus glands in frog ectoderm transplanted to a salamander embryo. The host is a salamander embryo with frog tissue on the ventral side of the head. Mucus glands were induced in the frog tissue and mucus is being secreted by them.

This case is of interest, since it shows that the salamander embryo can induce structures which are not a part of its own morphology. One is left with the impression that organizers are general stimuli, and that the end result is due to the tissue reacting according to the biological limitations of the species from which it is obtained.

RECONSIDERATIONS AND CONCLUSIONS

The organizer concept has shed some light on the old problem of mosaic and regulative eggs. In many cases, and perhaps in most, the change from an undetermined state to a determined state is the result of an influence external to the cells being determined. In other words, something of the nature of organizer action is involved. The main difference between mosaic and regulative eggs is thought to be the time at which these external influences cause determination. In the mosaic eggs this occurs during the time the egg is being formed in the ovary. The developing embryo is, therefore, mosaic from the start. In the parts of regulative eggs, on the other hand, determination occurs during the course of development.

Modern work in genetics and embryology has put the old controversy of preformation and epigenesis in a new light. Somewhat earlier, we saw that something 'preformed' must be transmitted from adult to embryo, otherwise the frog embryo might form a toad, a fish, or an elephant. Observations on developing embryos ruled out the transmission of preformed adult structures. Genetics and embryology have demonstrated that the preformed entities which are transmitted are the genes and an organized cytoplasm. A frog embryo becomes a frog because it receives the genes and cytoplasm that control the development of a frog body. The hereditary basis of development is preformed in the structure of the gametes; the appearance of adult parts is epigenetic.

This brief survey has revealed something of our vast ignorance concerning the factors responsible for differentiation. Embryology has not reached the point where we can say that we know why an embryonic cell differentiates in a certain way. We know some of the answers. When we know more we should be able to answer a question of vital interest: 'Why do normal cells sometimes change into cancerous cells?' This also is differentiation.

Further progress in embryology will depend upon knowledge in two closely related fields: (1) the manner in which genes act in development; and (2) cell physiology. Both of these fields are at the begining of their development.

In the next chapter we shall explore some of the ways in which the recent data of genetics can be applied to embryological problems.

SUGGESTED READINGS

Barth, L. G. 1953. *Embryology.* Dryden Press.

Brachet, J. 1950. *Chemical Embryology.* Interscience Publishers.

Cole, F. J. 1930. *Early Theories of Sexual Generation.* Oxford.

Huxley, J. S. and G. R. de Beer. 1934. *The Elements of Experimental Embryology.* Cambridge.

Needham, J. 1931. *Chemical Embryology.* Cambridge.

Needham, J. 1934. *History of Embryology.* Cambridge.

Needham, J. 1942. *Biochemistry and Morphogenesis.* Cambridge.

Rugh, R. 1948. *Experimental Embryology.* Burgess Publishing Co.

Spemann, H. 1938. *Embryonic Development and Induction.* Yale University Press.

Weiss, P. 1939. *Principles of Development.* Holt.

Wilson, E. B. 1937. *The Cell in Development and Heredity.* Macmillan. Chapters 13 and 14.

22

Developmental Control of Genetic Systems

The differentiation of embryonic cells must have as its basis the differentiation of the cell's gentoypes. This statement must come as a surprise when one recalls a fundamental principle of genetics: mitotic cell division produces daughter cells with genetic systems identical with those of the parent cell. While this principle is probably true, the supporting data, listed below, are not absolutely convincing.

1. The data of cytology show that the chromosomes of daughter cells are identical in number, structure, and staining characteristics with those of the parent cell.

2. There are many nuclear divisions between the zygote and the adult of a species (about 35 in the frog, for example). In all the the nuclear divisions of those cells in the lineage of ova or sperm, genetic integrity is maintained. We are sure of this because the genetic systems of ova and sperm can be tested by uniting them and studying the phenotype of the new individual. Except in exceptional circumstances, we cannot test as rigorously the genetic systems of differentiated somatic cells.

3. In many instances at least some of the somatic cells possess the same genotype as was present in the zygote. Thus, small parts of a Hydra or of a planarian worm can regenerate an entire individual. Some of the cells of the part, therefore, must have the complete genetic information of the species.

Let us consider these facts in relation to the discussion of Chapter 18. The data for the genetic control of protein synthesis are convincing. The DNA code is transmitted by the messenger RNA to the ribosomes. This gives the ribosome a specific surface configuration that becomes the basis for the synthesis of specific proteins. One message from DNA leads to the synthesis of hemoglobin; another message leads to the synthesis of insulin.

This explanation, involving DNA, messenger RNA, and ribosomes, is probably correct but, when considered from other points of view, some exceedingly difficult questions arise. In man and other vertebrates, hemoglobin is synthesized only in the cells that are about to become red blood cells. Similarly, insulin is synthesized only by the islet cells of the pancreas. So far as synthetic abilities are concerned, the red blood cells are differentiated one way and the islet cells another way. The differentiation of these two cell types begins in the early embryo: red blood cells arise from the mesoderm and islet cells from the endoderm. As the embryo develops, the two types of cells become increasingly different morphologically and finally differ in the specific proteins that they synthesize.

How can one explain the origin of these two cell types? One hypothesis might be: genes for hemoglobin synthesis are present in red blood cells but are absent from islet cells; genes for insulin synthesis are present in islet cells but are absent from red blood cells. Such a hypothesis holds that that there is a genetic difference between blood cells and islet cells—a difference reflected in their microscopic appearance and in their unique proteins.

This hypothesis is clearly at variance with the data discussed in the first part of the chapter, that suggest there is an identical genetic system in all an individual's cells. There are two main ways of resolving the dilemma.

First, we could assume that the data suggesting the genetic identity of all differentiated cells are inadequate and that the cells are, in fact, genetically different. Possibly only those cells that will form ova or sperm maintain intact the entire genetic system of the individual. Each type of somatic cell can then be thought of as genetically different from all other types. Evidence of the genetic nature of somatic cells is nearly always indirect (we must remember that only the genetics of germ cells can be tested adequately). Some evidence, such as regeneration in Hydra and planarian worms, suggests that at least some somatic cells have the full genetic system of the individual. Other evidence, such as the differing synthetic abilities of blood cells and islet cells, suggests that somatic cells may be genetically different.

Second, we could maintain that all cells, somatic and germ, have the same genetic systems, but that not all genes are active in all cell types or at all times. On the basis of this hypothesis, we would assume that islet cells possess genes responsible for both hemoglobin synthesis and insulin synthesis, but that only those concerned with insulin synthesis are active. The genes controlling hemoglobin synthesis are somehow inhibited in these cells. Furthermore, we might think of this inhibition as being either irreversible or reversible. If irreversible, it would be impossible for the insulin-synthesizing cells of the islets to produce hemoglobin under any experimental conditions. If, on the other hand, the inhibition were removed by some experimental condition hemoglobin might be produced.

There are many data that bear on these possibilities, but not enough for us to reach unequivocal conclusions. Some of these will now be considered.

Tests of Genetic Identity of Somatic Cells. Since the days of Roux, every student of development has been interested in the nature of the genetic system of somatic cells. Spemann was able to show that a single nucleus from a salamander embryo in the 16-cell stage, plus some cytoplasm, was able to develop into an entire embryo. Technical problems prevented his testing the developmental potentialities of older nuclei.

In 1952 R. Briggs and T. J. King (of the Institute for Cancer Research in Philadelphia) perfected a method for testing the nuclei of blastula and even older embryonic stages (Fig. 22–1). Basically their

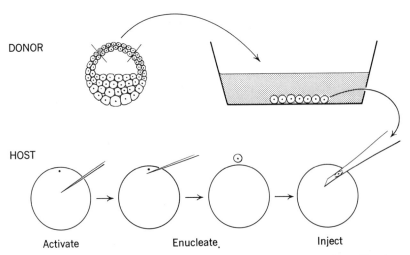

22–1 The Briggs and King method of transferring nuclei. A piece of an older donor embryo, in this case the roof of the blastocoel of a blastula, is removed and placed in a solution that causes the cells to fall free of one another. One cell is drawn into a micropipette and then injected into the host embryo. The host was previously prepared as follows: first, it was activated by being pricked with a glass needle; second, the egg nucleus was removed by flicking it out with a glass needle.

method consists of transferring a nucleus from older embryos to an unfertilized ovum from which the egg nucleus has been removed. The nucleus to be transferred is obtained as follows. In their first experiments, Briggs and King used embryos in the middle blastula stage (somewhat older than the embryo shown in Fig. 20–1). A portion of the blastocoel roof was cut off and a single cell from its underside drawn into a micropipette. Because the pipette had a bore smaller than the diameter of the cell, the cell must break as it enters the pipette. The nucleus, however, remains intact. The broken cell with its intact nucleus is then injected into the enucleated ovum. The injected ovum may now divide and form an embryo.

The experiment is difficult for both the experimenter and the embryo. Not infrequently the latter is injured, and either dies or develops abnormally. Other embryos, however, develop normally, and from these one can conclude that the transferred nucleus possesses all the information required for normal development. The injected nucleus of such an embryo would seem to be genetically identical to the zygote nucleus. Nuclei from cells of various parts of the blastula were tested in this manner. Briggs and King could find no evidence of any nuclear differentiation. These results support the hypothesis of the genetic identity of all the nuclei of an individual.

Older embryos were studied and different results were obtained. The evidence here suggests that the endodermal nuclei of late gastrulae and older embryos have undergone some sort of irreversible differentiation. When such nuclei are transplanted into enucleated ova, abnormal embryos result. The abnormalities are specific: the embryos have well-developed endodermal and mesodermal structures, but the ectodermal structures are strikingly deficient (the nervous system and sense organs are very poorly developed and sometimes the epidermis is so incomplete that it scarcely covers the embryo). Possibly these are the sorts of defects that one should expect if the injected nuclei have become differentiated in an 'endodermal direction.'

These results support the hypothesis that nuclei may change during the course of development. An endodermal nucleus of a blastula stage has the full genetic information of the zygote nucleus and, when injected into an enucleated ovum, can participate in fully normal development. By the late gastrula stage these same nuclei seem to have changed. Some of their potentialities, such as those necessary for the normal development of the ectodermal derivatives, are either lost or masked.

Briggs and King carried out their experiments with the frog, *Rana pipiens*. Similar work has been done by others using the African clawed frog, *Xenopus laevis*. The results obtained with Xenopus differ in important ways from those obtained with Rana. There is no clear evidence of nuclear differentiation in Xenopus. Nuclei from cells that have already undergone considerable visible differentiation can be injected into enucleated ova, and normal embryos, or even adults, may be obtained.

Thus the important question, whether or not nuclei become irreversibly differentiated during development, remains open. One must keep in mind, however, that the experiments test only for the presence of irreversible differentiation. It is possible that nuclei might become differentiated during development (namely, that genes may be inactivated in some cells but not in others); yet these same nuclei recover the full genetic potentialities of the species when returned to the biochemical environment of the uncleaved ovum.

A few pages back we spoke of a dilemma: the difficulties of reconciling the conflicting data and hypotheses of classical genetics and embryology. One possibility of solving the dilemma was to assume that the belief in the identity of all of the individual's nuclei might be incorrect. We have just shown that it is not possible to decide this question one way or another. The second possible solution is to assume that the genes are the same in all cells but that not all genes are active in all cells or at all times. Some recent data that support this hypothesis will now be cited.

Non-nuclear Control of Gene Action. Most geneticists of the first third of the twentieth century probably looked upon genes as controlling the cell's activities in ways quite uninfluenced by the cell as a whole. To be sure, genes required the substance of the cytoplasm for their work but this substance was something to be molded by the genes, not something to mold them. The data of genetics were most readily interpreted in this manner. Such an interpretation, however, was quite unacceptable to experimental embryologists. They were impressed by the importance of cytoplasmic materials. In mosaic eggs, especially, the pattern of development was associated with visible cytoplasmic differences (page 214). The gray crescent, a region of the egg's surface, seemed to be causally related to the blastopore (page 212). If an embryo was cut in half, the half with the gray crescent alone would form a blastopore and, later, a normal embryo. Thus a non-nuclear part of the embryo, not the nucleus, determined the outcome.

In recent years, many dramatic examples of the regulation of gene action by non-nuclear substances have been discovered. Most of these come from microorganisms, which have replaced Drosophila, corn, and other complex animals and plants as the materials most fruitful for gaining new insights into genetic phenomena.

In the bacterium *Escherichia coli* there is one gene that controls the synthesis of the enzyme, trytophane synthetase, and another that controls the synthesis of the enzyme, β-galactosidase. Presumably the mechanism is the usual one: the tryptophane synthetase gene transmits through messenger RNA to the ribosome the code for joining amino acids in the specific way that makes tryptophane synthetase. The events beginning with the β-galactosidase gene and ending with the formation of the specific enzyme β-galactosidase are apparently the same.

Tryptophane synthetase has a specific role in the cell. Under its catalytic influence, indole and serine are combined to form tryptophane. The amino acid tryptophane is one of the 20 found in all living organisms.

If *E. coli* cells are grown in a medium containing tryptophane, the enzyme is no longer synthesized. Thus the gene functions when there is no tryptophane, but is inhibited when tryptophane is present. This is a clear example of a gene's action being controlled by a non-nuclear substance.

The situation with respect to the β-galactosidase gene is almost the reverse. *E. coli* can use both glucose and galactose as a carbon source for intracellular syntheses. When there is no galactose in the cell's environment, no β-galactosidase is synthesized. If galactose is added to the medium in which the cells are growing, synthesis of the enzyme begins. We must assume that the gene does not function in the absence of galactose but functions when galactose is present. Once again, this is an example of non-nuclear control of a gene's activity.

Data such as these have led geneticists to a new hypothesis of gene action. One aspect of the hypothesis, which will not be developed in its entirety, is that there are different kinds of genes and chromosomal elements. We have discussed up to this point the 'structural genes' that produce messenger RNA. These are thought to be under the influence of an 'operator,' which, in turn, is under the control of a 'regulator gene.' The functional interrelations of the three sorts of genes in the hierarchy are thought to be as follows: The regulator gene produces a substance (possibly RNA) that represses the action of the operator. The operator, when so repressed, cannot stimulate the structural gene to act. Thus the structural gene cannot produce messenger RNA, without which there can be no protein synthesis (Fig. 22–2). This chain of inhibition is thought to be broken by certain 'substances' in the cytoplasm. These substances block the repressors produced by the regulator genes. If this occurs, the operator is not repressed and it stimu-

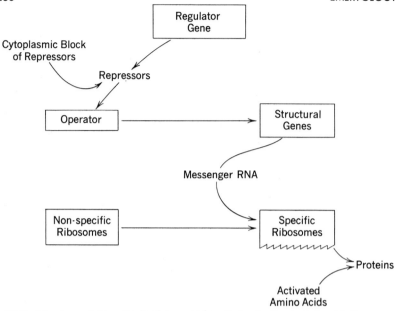

22-2   The current hypothesis of gene action. Refer to the text for details.

lates the structural gene. This, in turn, then produces messenger RNA and proteins can be synthesized.

There are many data supporting this hypothesis of gene action, which has been developed largely by F. Jacob and J. Monod of the Pasteur Institute in Paris. Not only does it satisfactorily account for many genetic facts, but it also provides an obvious way of explaining the role of genes in early development. The essential fact for embryologists is that non-chromosomal elements and conditions can be of paramount importance in controlling the action of structural genes and, ultimately, of the synthesis of specific proteins.

This hypothesis is consistent with the thinking of embryologists who fail to see how a genetic system, identical in all cells, alone provides for cellular differentiation. They conceive, instead, that external conditions or cytoplasmic substances interact with a uniform cellular genetic system to provide for differentiation. Though the genetic system specifies what a cell may do, non-genetic phenomena influence what it actually does. This point of view, which once would have been reasonable to an embryologist but not to a geneticist, now seems reasonable to both.

With this theoretical background, we will now recount one of the more striking examples from embryology of the influence of non-nuclear conditions on the course of cellular differentiation.

Cytodifferentiation in Fucus. Cellular differentiation in the seaweed Fucus (commonly called rockweed) is already present at the first possible opportunity, that is, by the two-cell stage. The zygote is at first spherical (Fig. 22–3). Before first cleavage begins, a protuberance appears on one side, giving the zygote roughly the shape of a snow-shoe. The polarity thus developed is the basis of further differentiation.

First cleavage occurs across the long axis of the zygote, producing two cells of different shapes—only one having the protuberance. The developmental fates of the two cells are fundamentally different. The cell with the protuberance divides repeatedly to give rise to the rhizoid, which attaches the Fucus to the rocks. The other cell gives rise to the thallus, which is the leaf-like part of the plant. According to D. M. Whittaker, now of The Rockefeller Institute, who made these observations on Fucus, 'when the point of origin of the rhizoid protuberance is determined, the polarity and whole developmental pattern of the embryo is determined.'

The factors responsible for the formation of the protuberance, therefore, are of fundamental importance in determining the differentiation of cells in Fucus. Whittaker found that seemingly minor environmental differences could determine the point of outgrowth of the protuberance. Consequently, if the zygotes are in a group, the protuberances are directed inward (Fig. 22–3). If the zygote is placed between two pipettes, one with seawater agar at pH 7.8 and the other at pH 6.4, the protuberance appears on the side of the lower pH. If the zygotes are exposed to white light, the protuberance appears on the dark side. If the zygotes are kept in a temperature gradient, the protuberance appears on the warmer side.

Thus an initial stimulus, which has nothing to do with the nucleus, controls the beginning of a series of events that is of profound importance in cellular differentiation. The protuberance begins when the embryo contains a single nucleus. The first division occurs some hours after the protuberance has been induced, and in a plane related to the protuberance itself. Recall also that, once the protuberance has formed, the basic polarity of the developing embryo has been determined. Any specific differentiation of the nucleus that may occur later is, therefore, the result of an initial environmental stimulus.

There is much evidence from many other organisms suggesting that the basic polarity of an egg, hence of the embryo itself, is determined by conditions external to the maturing egg cell. In these important events, the egg's nucleus seems to play no determining role. In many mosaic embryos the pattern of development is closely correlated with

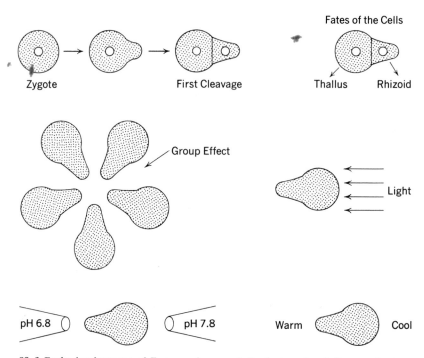

22-3 Early development of Fucus and some of the factors that influence the formation of the protuberance.

the distribution, during cleavage, of visibly different cytoplasmic regions.

The establishment of some polarity or regional differences in an embryo is an exceedingly important factor in development. Consider the result if this were not the case: a nucleus dividing by mitosis in an unorganized cytoplasm would produce only a group of similar cells, certainly not a differentiated embryo. There must be some initial induction of a difference, some defining of polarity. Once this has been brought about, one can suggest models for the manner in which cellular differentiation can proceed, even with nuclei that are at first functionally identical. The experiments on the formation of tryptophane synthetase and β-galactosidase in *E. coli* immediately come to mind. But the initial functional identity of the nuclei must be short-lived because, in the course of development, different cells begin to produce different proteins. It is probably legitimate to conclude that, if some of the proteins of two different cell types are different, then some of the genes of the two types are functioning differently. In this general sense we might say that genetic differentiation had occurred.

The genes that would otherwise control the synthesis of hemoglobin are presumably inactivated by the cytoplasmic environment of all cells other than red blood cells.

Cellular differentiation may be thought of as an interaction between nuclei, which in the beginning at least are identical, and different cytoplasmic regions. In many cases these cytoplasmic regions are formed under the control of genes that acted much earlier, usually when the ova were being formed. In other cases, such as Fucus, the environment may be important in inducing cytoplasmic differences. Possibly the substances that give the cytoplasm its regional specificity interact with the repressor substances (Fig. 22–2) produced by the regulator genes. Whatever the mechanism may be, it seems probable that nuclear differentiation is concomitant with the differentiation of the cell as a whole. A nucleus of an islet cell does not develop the way it does because of some innate specificity. Instead it *is developed*—developed in a specific way because of the cytoplasm in which it happens to lie.

A generation ago few embryologists or geneticists would have predicted that a synthesis of their fields would be made possible by studies on the bacterium *Escherichia coli*. But this microscopic creature, with no embryology of its own, has shown a way. A decade from now it may be difficult to distinguish between a geneticist and an embryologist, as they advance their science beyond what each might independently achieve.

SUGGESTED READINGS

Jacob, F. and J. Monod. 1961. 'On the regulation of gene activity.' *Cold Spring Harbor Symposia on Quantitative Biology 26*:193–209.

Moore, J. A. 1962. 'Nuclear transplantations and problems of specificity in developing embryos.' *Journal of Cellular and Comparative Physiology*, Suppl. 1, *60*:19–34. (The Fucus example is taken from this article.)

Sussman, M. 1960. *Animal Growth and Development*. Prentice-Hall.

Waddington, C. H. 1956. *Principles of Embryology*. George Allen & Unwin.

Index